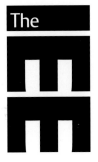

Code of Practice for In-Service Inspection and Testing of Electrical Equipment

3rd Edition

Published by The Institution of Engineering and Technology, London, UK

© 1994, 2001, 2007 The Institution of Engineering and Technology

First published 1994 (0 85296 844 2)
Second edition 2001 (0 85296 776 4)
Reprinted with new cover 2003
Third edition 2007 (978-0-86341-833-4)
Reprinted 2008

The Institution of Engineering and Technology is the new institution formed by the joining together of the IEE (The Institution of Electrical Engineers) and the IIE (The Institution of Incorporated Engineers). The new Institution is the inheritor of the IEE brand and all its products and services, such as this one, which we hope you will find useful.

Copies may be obtained from:
The Institution of Engineering and Technology
PO Box 96
Stevenage
SG1 2SD, UK
Tel: +44 (0)1438 767328
Email: sales@theiet.org
www.theiet.org/publishing/books/wir-reg/

ISBN 978-0-86341-833-4

Contents

Code of Practice for In-Service Inspection and Testing of Electrical Equipment
© The Institution of Engineering and Technology

Cooperating organisations

The Institution of Engineering and Technology wishes to acknowledge the contribution made by the following representatives of organisations in the preparation of this Code of Practice.

Association of Manufacturers of Domestic Electrical Appliances (AMDEA)
Mr S. MacConnacher

British Cables Association (BCA)
Mr C. Reed

British Electrotechnical and Allied Manufacturers Association (BEAMA)
Mr P. Galbraith
Mr R. Lewington
Mr M. Mullins
Mr P. Sayer

Benchmark Electrical Safety Technology Ltd
Mr G. Stokes

British Electrotechnical Approvals Board (BEAB)

City & Guilds
Ms J. Andrews
Mr R. Mansell
Mr R. Woodcock

CORGI
Mr P. Collins

Electrical Contractors' Association (ECA)
Mr D. Locke
Mr H. R. Lovegrove
Mr L. Markwell

Electrical Safety Council

Energy Networks Association (ENA)
Mr D. Start

ERA
Mr M. Coates

GAMBICA Association Ltd
Mr M. Johnson
Mr P. Lawson
Mr K. Morriss
Mr A. Sehdev
Mr J. Wallace

Health and Safety Executive (HSE)
Mr K. Morton

Electrical Contractors' Association of Scotland (SELECT)
Mr D. Millar

Institution of Engineering and Technology (IET)
Mr C. Buck
Mr N. Canty
Mr M. Coles
Mr G. Cronshaw
Mr N. Friswell
Mrs S. Jupp
Mr I. Turner
Mr G. Willard

NICEIC
Mr F. Skipper

POLYCOM
Mr J. Pearson

SJCF
Mr R. Van der Post

Society of Electrical and Mechanical Engineers serving Local Government (SCEME)
Mr C. Tanswell

Author
Mr J. Ware

Acknowledgements

References to British Standards are made with the kind permission of BSI. Complete copies can be obtained by post from:

BSI Customer Services
389 Chiswick High Road
London W4 4AL
Tel: +44 (0)20 8996 9001
Fax: +44 (0)20 8996 7001
Email: orders@bsi-global.com

References to HSE publications are made with the kind permission of the HSE. Copies of publications can be obtained from:

HSE Books
PO Box 1999
Sudbury
Suffolk CO10 2WA
Tel: +44 (0)1787 881165
Email: hsebooks@prolog.uk.com
Web: www.hsebooks.com

HSE publications are also available through good booksellers.

Further information is available from:

HSE Infoline: +44 (0)845 345 0055
Email: hseinformationservices@natbrit.com
Web: www.hse.gov.uk

Preface

The objective of this Code of Practice is to give advice on in-service inspection and testing to determine whether electrical equipment is fit for continued service or maintenance or replacement is necessary.

The main changes in this edition of the Code include clarification of which equipment is covered, additional explanation of the term 'competent person' and inclusion of the substitute/alternative test method. In addition, there is explanation on the use of test leads, and RCD adaptors and extension leads are now covered.

As many deficiencies result from the plug or the cable, new appendices give a series of pictures, which illustrate commonly encountered deficiencies.

The Code recognizes the imminent publication of the 17th Edition of BS 7671: 2008.

Code of Practice for In-Service Inspection and Testing of Electrical Equipment
© The Institution of Engineering and Technology

Introduction

Why is it necessary to maintain electrical equipment?

The Electricity at Work Regulations require, in Regulation 4(2), that:

> As may be necessary to prevent danger, all systems shall be maintained so as to prevent, so far as is reasonably practicable, such danger.

Regulation 4(2) is concerned with the need for maintenance to be done in order to ensure safety of the system if danger would otherwise result. The quality and frequency of maintenance should be sufficient to prevent danger so far as is reasonably practicable.

Regular inspection of equipment is an essential part of any preventive maintenance programme. Practical experience of the use of equipment and the conditions prevailing may indicate an adjustment to the frequency at which preventive maintenance needs to be carried out. This is a matter for the judgement of the responsible person, who should seek all the information needed to make the judgement including reference to the manufacturer's guidance.

Records of maintenance, including test results, should be kept throughout the working life of the electrical equipment to enable

- ► the condition of the equipment to be monitored,
- ► the effectiveness of the maintenance policies to be assessed, and
- ► to demonstrate that an effective maintenance system is in place.

What should be maintained?

All electrical systems and equipment should be maintained if danger would otherwise arise (see Figure 1).

Other than the fixed installation, all electrical equipment in an installation, whether permanently connected or connected by a plug and socket-outlet, should be inspected and tested in accordance with the recommendations contained in this Code of Practice.

The fixed installation should be periodically inspected and tested to ensure satisfactory condition for continued use as required by BS 7671: *Requirements for Electrical Installations*. Guidance on the requirements contained in BS 7671 concerning inspection and testing of the fixed electrical installation is given in Guidance Note 3: *Inspection and Testing*.

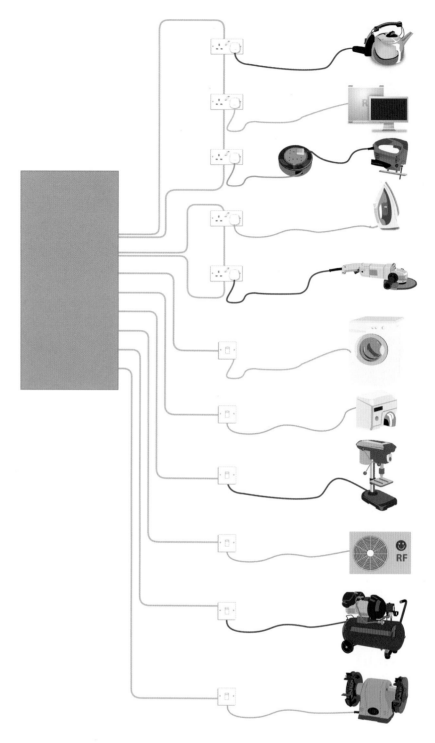

▼ Figure 1
This Code of Practice provides guidance on the inspection and testing of: portable, movable and hand-held tools and equipment; equipment connected by means of a cable or cord connected to an outlet plate; appliances to build in; IT equipment; extension leads, multiway adaptors and suppressor adapters

Who has responsibilities?

The following people have responsibility for electrical systems and equipment.

▶ Users of electrical equipment (whose responsibilities include user checks).
▶ Administrators with responsibility for electrical maintenance who may not necessarily have detailed technical knowledge.
▶ The competent person carrying out the formal visual inspection and the inspections and tests.
▶ Other dutyholders such as company directors, managers or building services managers.

What needs to be done to comply with the relevant requirements of the Electricity at Work Regulations?

The requirements of the Electricity at Work Regulations can be met by

1   performing in-service inspection and testing, which consist of three activities:
    i   user checks
    ii formal visual inspections (without tests)
    iii combined inspections and tests
2   performing maintenance or, if necessary, replacing the defective item of equipment (depending upon the results of the in-service inspection and testing), and
3   keeping up-to-date records that can be a means of showing compliance.

Information on the Electricity at Work Regulations can be found in the HSE publication *The Memorandum of Guidance on the Electricity at Work Regulations 1989* (see Figure 2).

▼ **Figure 2**
HSR 25: Memorandum of Guidance on the Electricity at Work Regulations 1989 [Courtesy of the HSE]

Background to the Code of Practice

To encourage free trade within the European Union, existing national standards are being harmonized and converted to European standards. Compliance with harmonized European standards gives assurance to purchasers that appliances and equipment have been designed and constructed to a standard that will ensure that in normal use they function safely and without danger.

In order to check compliance, manufacturers have to perform a series of tests on the appliance and its components as required by the standard. The appliance is required to pass these tests if it is to be said that the appliance complies with the standard. A list of some of the safety standards for electrical equipment is given in Appendix I. The tests detailed in these standards are generally not suitable for in-service testing.

This Code of Practice recommends in-service inspections and tests that can be applied generally to equipment and appliances in normal use. Routine manufacturers' tests are not required for general in-service testing, but may be applied to new appliances or after repair. Further information is provided in Appendix V.

Layout of the Code of Practice

Part 1 provides guidance on what work should be done in order to comply with the applicable legislation, including the Electricity at Work Regulations, and whether this work can be carried out in-house. Advice is included on the law, procedures, documentation and training.

Part 2 is written for those carrying out the practical work and explains the details of the inspections and tests.

Part 3 comprises a series of appendices containing information and guidance and includes model forms that allow records to be kept in order to demonstrate that an effective system of maintenance is in place.

Code of Practice for In-Service Inspection and Testing of Electrical Equipment

Part 1
Administration of Inspection and Testing

Scope

1.1 Users of electrical equipment, persons managing a maintenance scheme, persons performing inspections and tests and other dutyholders

This Code of Practice gives guidance for:

1 Users of electrical equipment.
2 Persons with administrative responsibilities for maintenance of electrical equipment (Administrators), either within or contracted to an organisation or authority, such as facility managers. Part 1 of this Code provides information for persons with these particular responsibilities.
3 Persons undertaking the practical inspection and testing of electrical equipment. Such persons should study and become fully conversant with all of this Code. Part 2 gives detailed information on testing that will need to be referred to frequently.
4 Other dutyholders such as company directors, managers and building service managers.

1.1.1 Users of electrical equipment

Users of electrical equipment have responsibilities, which include ensuring the equipment they use has no obvious visual damage or defects. Users should be instructed to remove from service and report any defective equipment. Detailed checks that the user should perform before using an item of equipment are given in Chapter 13 of this publication.

1.1.2 Persons managing a maintenance scheme (Administrators)

Advice to administrators in three areas is given in Part 1 of this publication.

▶ The issues relating to combined inspection and testing are discussed so that administrators may make decisions concerning arrangements for inspection and testing, including the frequency of the combined inspections and tests.
▶ An explanation is given of the information that service administrators should receive from any organisation carrying out combined inspection and testing.
▶ Information is provided so that administrators can decide whether the combined inspection and testing could be carried out in-house.

1.1.3 Persons undertaking the practical inspection and testing of electrical equipment

Detailed information on the inspections and tests necessary is given to competent persons who have specific responsibilities for carrying out the formal visual inspections and the combined electrical inspections and tests, either in-house or on a contractual basis for a client.

1.1.4 Other dutyholders such as company directors, managers and building services managers

The guidance in this Code of Practice is intended to assist dutyholders in meeting the requirements of the Electricity at Work Regulations. It sets out the Regulations and gives guidance on them. Although it reflects the IET's view of the meaning of terms used in the Regulations, only the Courts can provide a binding interpretation. The purpose of this Code of Practice is to amplify the nature of the precautions in general terms so as to help in the achievement of high standards of electrical safety in compliance with the duties imposed.

1.2 Equipment

The equipment covered by this Code of Practice includes Class I, Class II and Class III equipment of the following types:

- ▶ portable equipment
- ▶ movable equipment
- ▶ stationary equipment
- ▶ hand-held equipment
- ▶ equipment that is plugged in
- ▶ equipment connected by means of a flexible cord or cable to a fused or unfused connection unit or isolator
- ▶ built-in appliances
- ▶ IT equipment
- ▶ extension leads, RCD extension leads, multiway adaptors, RCD adaptors
- ▶ equipment with high protective conductor currents

1.3 Premises

This Code of Practice can be applied to electrical equipment in industrial, commercial and domestic environments. It applies to all premises including offices, shops, hotels, schools, universities, hospitals, theatres, museums, general industrial locations, construction sites, portable buildings, caravans (both static and mobile), swimming pools and agricultural and horticultural premises. It can be used for the inspection and testing of equipment in rented properties.

1.4 Voltages and phases

This Code applies to equipment supplied at voltages up to and including 1000 V a.c. or 1500 V d.c. between conductors or 600 V a.c. or 900 V d.c. between conductors and Earth including single-, two- and three-phase equipment supplied at 400 V, 230 V and 110 V and at extra-low voltage including SELV (Separated Extra-Low Voltage).

1.5 Summary of the objectives of this Code of Practice

The objective of this Code of Practice is to give advice on in-service inspection and testing to determine whether electrical equipment is fit for continued service or maintenance or replacement is necessary (Figure 1.1). Inspection and testing, subsequent maintenance (if needed) and adequate record keeping performed in accordance with the guidance given in this Code of Practice are likely to achieve conformity with the relevant parts of the Electricity at Work Regulations.

The fixed electrical installation to which the equipment is connected should not be faulty and brief reference is made to the initial and periodic testing of the fixed installation.

This Code does not deal with the legislation relating to the supply of equipment, whether new or secondhand, to a third party by way of sale, hire or other method.

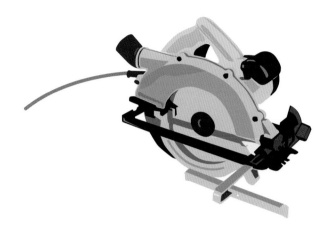

▼ **Figure 1.1**
Electrical equipment must be fit for continued service

© The Institution of Engineering and Technology

Definitions

(The definitions given are from BS 7671: 2001(2004) *Requirements for Electrical Installations* except for those marked[1]. Note that some definitions may change in BS 7671: 2008.)

Accessory. A device, other than current-using equipment, associated with such equipment or with the wiring of an installation.

Appliance. An item of current-using equipment other than a luminaire or an independent motor.

Appliances or equipment for building-in[1]**.** Equipment intended to be installed in a prepared recess such as a cupboard or similar. In general, equipment for building-in does not have an enclosure on all sides.

Basic insulation. Insulation applied to live parts to provide basic protection against electric shock and which does not necessarily include insulation used exclusively for functional purposes.

Basic protection[1]**.** Protection against electric shock under fault-free conditions. (Previously referred to as protection against direct contact.)
Note: This definition is planned for inclusion in BS 7671: 2008.

Circuit protective conductor (cpc). A protective conductor connecting exposed-conductive-parts of equipment to the main earthing terminal.

Class 0 equipment[1]**.** Equipment in which protection against electric shock relies upon basic insulation only. There is no supplementary or reinforced insulation. There is no provision for the connection of accessible conductive parts, if any, to the protective conductor in the fixed wiring of the installation. In the event of failure of the basic insulation reliance is placed upon the environment in which the equipment is installed.
Note: This definition is from BS 2754: *Memorandum. Construction of electrical equipment for protection against electric shock.*

Class I equipment. Equipment in which protection against electric shock does not rely on basic insulation only, but which includes means for the connection of exposed-conductive-parts to a protective conductor in the fixed wiring of the installation. Figure 2.1 illustrates a Class I appliance. (Figure 2.3 illustrates an appliance that may be Class I or Class II.) Refer also to BS 2754.

▼ Figure 2.1
Most bench grinders
are Class I items of
equipment and have a
metal case that should
be reliably connected
with Earth

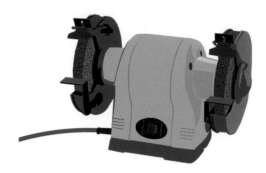

Class II equipment. Equipment in which protection against electric shock does not rely on basic insulation only, but in which additional safety precautions such as supplementary insulation are provided, there being no provision for the connection of exposed metalwork of the equipment to a protective conductor and no reliance upon precautions to be taken in the fixed wiring of the installation. Figure 2.2 illustrates a Class II appliance. Refer also to BS 2754.

▼ Figure 2.2 Hairdryers are usually
Class II appliances. Protection against electric
shock does not rely on basic insulation only,
but additional safety precautions such as
supplementary insulation are provided, and
there is no provision for the connection
of any exposed metalwork to a protective
conductor and no reliance is made upon
precautions taken in the fixed wiring of the
installation

▼ Figure 2.3
An angle grinder may
be a Class I item with
metal parts earthed
or it may be Class II. If
Class II, it should carry
the double insulated
symbol. See Figure
11.12. If it is not known
whether an item of
equipment is Class I
or Class II, it should be
treated as Class I

Class III equipment. Equipment in which protection against electric shock relies on a supply at SELV and in which voltages higher than those of SELV are not generated. Refer also to BS 2754.
Note: See definition of SELV.

Cord set[1]**.** An assembly consisting of a detachable flexible cable or cord fitted with a plug and a connector intended for the connection of electrical equipment to the electrical supply.

Competent person[1]. A person possessing sufficient technical knowledge or experience to be capable of ensuring that injury is prevented. Technical knowledge or experience may include

1 adequate knowledge of electricity,
2 adequate experience of electrical work,
3 adequate understanding of the system to be worked on and practical experience of that class of system,
4 understanding the hazards that may arise during the work and the precautions that need to be taken, or
5 the ability to recognize at all times whether it is safe for work to continue.

Refer to Regulation 16 of the Electricity at Work Regulations and the guidance included in HSE publication HSR 25.

Current-using equipment. Equipment that converts electrical energy into another form of energy, such as light, heat or motive power.

Danger. Risk of injury to persons (and livestock where expected to be present) from

1 fire, electric shock and burns arising from the use of electrical energy, and
2 mechanical movement of electrically controlled equipment, in so far as such danger is intended to be prevented by electrical emergency switching or by electrical switching for mechanical maintenance of non-electrical parts of such equipment.

Direct contact. Contact of persons or livestock with live parts (see Figures 2.4 and 2.5).
Note 1: See also Basic protection.
Note 2: Protection against direct contact is afforded by means such as insulation, barriers and enclosures and is referred to as basic protection.

Double insulation. Insulation comprising both basic insulation and supplementary insulation.

Earth. The conductive mass of the Earth, whose electric potential at any point is conventionally taken as zero.

▼ **Figure 2.4**   ▼ **Figure 2.5**

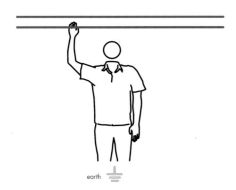

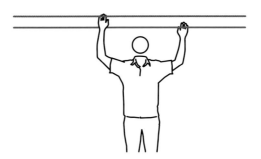

▼ **Figure 2.4**
Direct contact by contact with a live conductor while in contact with Earth

▼ **Figure 2.5**
Direct contact by contact with two live conductors at different potentials

Earthing. Connection of the exposed-conductive-parts of electrical equipment to the main earthing terminal of an electrical installation.

Note: Metal parts of an electrical installation or appliance may become electrically charged if there is a fault. The purpose of earthing is to minimize the risk of electric shock should anyone touch those metal parts when a fault is present. This is achieved by providing a path for fault current to flow safely to Earth, which also causes the protective device to operate and disconnect the circuit thereby removing the danger (see Figure 2.6).

Electric shock. A dangerous physiological effect resulting from the passing of an electric current through a human body or livestock.

Electrical equipment. (abbr: Equipment). Any item for such purposes as generation, conversion, transmission, distribution or utilization of electrical energy, such as machines, transformers, apparatus, measuring instruments, protective devices, wiring systems, accessories, appliances and luminaires.

▼ **Figure 2.6**
Earthing is achieved by providing a path for fault current to flow safely to Earth, which causes the protective device to operate and disconnect the supply thereby removing the danger

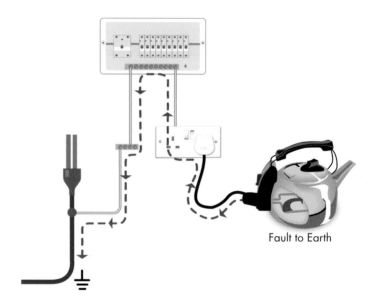

Fault to Earth

Electrical installation. (abbr: Installation). An assembly of associated electrical equipment supplied from a common origin to fulfil a specific purpose and having certain coordinated characteristics.

Enclosure. A part providing protection of equipment against certain external influences and in any direction protection against direct contact.

Exposed-conductive-part. A conductive part of equipment that can be touched and that is not a live part, but that may become live under fault conditions.

Extra-low voltage. See Voltage, nominal.

Fault. A circuit condition in which current flows through an abnormal or unintended path. This may result from an insulation failure or a bridging of insulation. Conventionally the impedance between live conductors or between live conductors and exposed-conductive-parts or extraneous-conductive-parts at the fault position is considered negligible.

Fault protection[1]. Protection against electric shock under single fault conditions. (Previously referred to as protection against indirect contact.)
Note: This definition is planned for inclusion in BS 7671: 2008.

Fixed equipment. Equipment designed to be fastened to a support or otherwise secured in a specific location.

Flexible cable. Cable whose structure and materials make it suitable to be flexed while in service.

Flexible cord. A flexible cable in which the cross-sectional area (csa) of each conductor does not exceed $4\,mm^2$.

Fuse. A device that, by the melting of one or more of its specially designed and proportioned components, opens the circuit in which it is inserted by breaking the current when this exceeds a given value for a sufficient time. The fuse comprises all the parts that form the complete device.

Hand-held appliance or equipment[1]. Portable equipment intended to be held in the hand during normal use, e.g. hairdryer, power drill, hedge-cutter, soldering iron.

Indirect contact. Contact of persons or livestock with exposed-conductive-parts that have become live under fault conditions.
Note 1: See Fault protection.
Note 2: Indirect contact occurs when a person is in contact with a conductive part, such as a metal enclosure, that has become live due to a fault. Protection against indirect contact is referred to as fault protection (see Figure 2.7).

Information technology equipment[1]. Information technology (IT) equipment includes electrical business equipment such as computers and mains powered telecommunications equipment, data network equipment, banking equipment and other equipment for general business use, such as electrical and electronic office equipment and retail equipment including mail processing machines, electric plotters, trimmers, VDUs, data terminal equipment, typewriters, telephones, printers, photocopiers, power packs and mobile telephone charging units.

Instructed person. A person adequately advised or supervised by a skilled person to enable him/her to avoid the dangers that electricity may create.

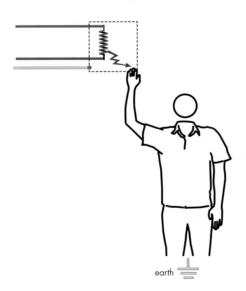

earth

▼ **Figure 2.7**
Indirect contact: Contact with a conductive part that has become live under fault conditions

Insulation. Suitable non-conductive material enclosing, surrounding or supporting a conductor.

IP (Index of Protection) code[1]**.** The IP code includes two numbers. The first number indicates the degree of protection of equipment within the enclosure against the ingress of solid foreign bodies. The second number indicates the degree of protection of equipment against the ingress of water (see Table 2.1). An example of an IP code would be IP34.

▼ **Table 2.1**
The first two digits of the IP code

Solid foreign bodies		Water	
0	Not protected	0	Not protected
1	Protected against solid foreign objects of 50 mm diameter or greater	1	Protected against vertically falling drops
2	Protected against solid foreign objects of 12.5 mm diameter or greater	2	Protected against vertically falling drops with the item tilted 15 degrees
3	Protected against solid foreign objects of 2.5 mm diameter or greater	3	Protected against spraying water
4	Protected against solid foreign objects of 1.0 mm diameter or greater	4	Protected against splashing water
5	Dust protected	5	Protected against water jets
6	Dust tight	6	Protected against powerful water jets
		7	Protected against temporary immersion
		8	Protected against continuous immersion

Additional letters are optional (such as IPXXB) and they indicate the degree of protection of persons against access to hazardous parts (see Table 2.2).

▼ **Table 2.2**
Additional letters used in the IP code

Additional letter	Brief description	Definition
A	Protected against access with the back of the hand	The access probe, a sphere of 50 mm diameter, is required to have adequate clearance from hazardous parts
B	Protected against access with a finger	The jointed test finger of 12 mm diameter and 80 mm length is required to have adequate clearance from hazardous parts
C	Protected against access with a tool	The access probe of 2.5 mm diameter and 100 mm length is required to have adequate clearance from hazardous parts
D	Protected against access with a wire	The access probe of 1 mm diameter and 100 mm length is required to have adequate clearance from hazardous parts

For further information refer to BS EN 60529: *Degrees of protection provided by enclosures (IP code)*.

Code of Practice for In-Service Inspection and Testing of Electrical Equipment
© The Institution of Engineering and Technology

Isolation. A function intended to cut off for reasons of safety the supply from all, or a discrete section, of the installation by separating the installation or section from every source of electrical energy.

Leakage current. Electric current in an unwanted conductive path under normal operating conditions.

Live part. A conductor or conductive part intended to be energized in normal use, including a neutral conductor but, by convention, not a PEN conductor.
Note: A PEN conductor is a conductor combining the functions of both protective conductor and neutral conductor.

Low voltage. See Voltage, nominal.

Luminaire. Equipment that distributes, filters or transforms the light from one or more lamps and that includes any parts necessary for supporting, fixing and protecting the lamps, but not the lamps themselves, and, where necessary, circuit auxiliaries together with the means for connecting them to the supply. For the purposes of the Regulations a lampholder, however supported, is deemed to be a luminaire.
Note: This definition may change in BS 7671: 2008.

Means of earthing[1]**.** A means of earthing is an arrangement through which a connection is made to the general mass of Earth for the exposed-conductive-parts of an installation. The exposed-conductive-parts are connected to the means of earthing via the earthing conductor and main earthing terminal of the installation.

Movable equipment[1] (sometimes called transportable). Equipment that is either

▶ 18 kg or less in mass and not fixed, e.g. electric fire, small welding set (Figure 2.8), or
▶ equipment with wheels, castors or other means to facilitate movement by the operator as required to perform its intended use, e.g. air-conditioning unit.

▼ **Figure 2.8**
A welding set will normally be considered as movable equipment

Nominal voltage. See Voltage, nominal.

Ordinary person. A person who is neither a skilled person nor an instructed person.

PAT instrument[1]. Portable appliance test instrument.

Portable appliance[1]. An appliance of less than 18 kg in mass that is intended to be moved while in operation or an appliance that can easily be moved from one place to another, e.g. kettle, toaster, food mixer, vacuum cleaner.

Portable equipment. Electrical equipment that is moved while in operation or which can easily be moved from one place to another while connected to the supply.

Protective conductor. A conductor used for some measures of protection against electric shock and intended for connecting together any of the following parts:

▶ exposed-conductive-parts
▶ extraneous-conductive-parts
▶ the main earthing terminal
▶ earth electrode(s)
▶ the earthed point of the source, or an artificial neutral

Protective conductor current. Electric current that flows in a protective conductor under normal operating conditions (see Figure 2.9).

▼ **Figure 2.9**
A protective conductor current will result due to the capacitors and discharge resistors in the EMC filter

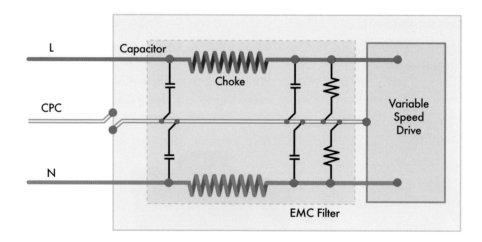

Reinforced insulation. Single insulation applied to live parts that provides a degree of protection against electric shock equivalent to double insulation under the conditions specified in the relevant standard. The term 'single insulation' does not imply that the insulation has to be in one homogeneous piece. It may comprise several layers that cannot be tested singly as supplementary or basic insulation.

RCD (Residual Current Device). A mechanical switching device or association of devices intended to cause the opening of the contacts when the residual current attains a given value under specified conditions.

SELV (Separated Extra-Low Voltage). An extra-low voltage system that is electrically separated from Earth and from other systems in such a way that a single fault cannot give rise to the risk of electric shock (see Figure 2.10).
Note: SELV is described as 'Safety' Extra-Low Voltage in some standards.

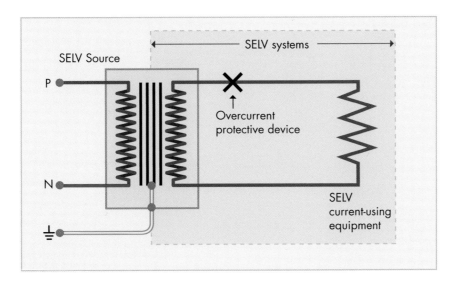

▼ **Figure 2.10** SELV system. The principal requirements for an a.c. SELV system are that the source is required to be a safety source, the voltage is not to exceed 50 V rms, the circuit conductors are to be physically separated from those of any other system or other arrangements made and no exposed-conductive-part of a SELV system is to be connected to Earth, to any other exposed-conductive-part, to a circuit protective conductor or to an extraneous-conductive-part unless other arrangements are made. Socket-outlets are not to have protective conductor contacts. An example of a safety source is a double wound transformer meeting the requirements of BS EN 60742: *Isolating transformers and safety isolating transformers. Requirements.*

Skilled person. A person with technical knowledge or sufficient experience to enable him/her to avoid the dangers that electricity may create.

Socket-outlet. A device, provided with female contacts, that is intended to be installed with the fixed wiring and intended to receive a plug. A luminaire track system is not regarded as a socket-outlet system.

Stationary equipment or appliance[1]. Electrical equipment that is either fixed, or equipment having a mass exceeding 18 kg and not provided with a carrying handle, e.g. refrigerator or washing machine.

Supplementary insulation. Independent insulation applied in addition to basic insulation in order to provide protection against electric shock in the event of a failure of basic insulation.

Touch current[1]. Electric current through a human body or through an animal's body when it touches one or more accessible parts of an installation or equipment.

Voltage, nominal. Voltage by which an installation (or part of an installation) is designated. The following ranges of nominal voltage (rms values for a.c.) are defined:

► Extra-low – normally not exceeding 50 V a.c. or 120 V ripple-free d.c., whether between conductors or to Earth.
► Low – normally exceeding extra-low voltage but not exceeding 1000 V a.c. or 1500 V d.c. between conductors, or 600 V a.c. or 900 V d.c. between conductors and Earth.

The actual voltage of the installation may differ from the nominal value by a quantity within normal tolerances.

The law

3

3.1 The legislation

Electrical equipment is required to be properly maintained so as to prevent danger. Inspections and tests are necessary. Although reference is made to legislation, this chapter should not be considered as legal advice. In case of doubt, the specific legislation mentioned should be consulted and legal advice obtained.

The responsibilities for safety of persons at work are detailed in legislation and some of the applicable legislation is listed in Appendix II.

The legislation relevant to electrical maintenance is:

- Health and Safety at Work etc. Act 1974
- Management of Health and Safety at Work Regulations 1999
- Provision and Use of Work Equipment Regulations 1998
- Electricity at Work Regulations 1989
- Workplace (Health, Safety and Welfare) Regulations 1992

3.1.1 The Health and Safety at Work etc. Act 1974

This Act puts a duty of care upon both employer (Sections 2, 3 and 4 etc.) and employee (Section 7) to ensure the safety of all persons using the work premises. This includes the self-employed.

3.1.2 The Management of Health and Safety at Work Regulations 1999

These regulations state:

> Every employer shall make a suitable and sufficient assessment of:
>
> **(a)** the risks to the health and safety of his employees to which they are exposed while they are at work, and
> **(b)** the risks to the health and safety of persons not in his employment arising out of or in connection with the conduct by him of his undertaking (Regulation 3(1)).

3.1.3 The Provision and Use of Work Equipment Regulations 1998

These regulations state:

> Every employer shall ensure that work equipment is so constructed or adapted as to be suitable for the purpose for which it is used or provided (Regulation 4(1)).

The Provision and Use of Work Equipment Regulations 1998 (PUWER) cover most risks that can result from using work equipment. With respect to risks from electricity, compliance with the Electricity at Work Regulations 1989 is likely to achieve compliance with PUWER Regulations 5–9, 19 and 22.

PUWER only applies to work equipment used by workers at work. This includes all work equipment (fixed, portable or transportable) connected to a source of electrical energy. PUWER does not apply to the fixed installations in a building. The electrical safety of these installations is just one of the issues dealt with by the Electricity at Work Regulations.

Regulation 4 of PUWER requires the employer to ensure that equipment is only used for operations and under conditions for which it is suitable.

3.1.4 The Electricity at Work Regulations 1989

These regulations apply to electrical equipment as defined in the Regulations, which includes every type of electrical equipment from, for example, a 400 kV overhead line to a battery-powered hand lamp. It is appropriate for the Regulations to apply even at the very lowest end of the voltage or power spectrum because the Regulations are concerned with, for example, explosion risks that may be caused by very low levels of energy igniting flammable gases even though there may be no risk of electric shock or burn. Thus no voltage limits appear in the Regulations. The criterion of application is the test as to whether 'danger' (as defined) may arise.

Electrical equipment (as defined) includes conductors used to distribute electrical energy such as cables, wires and leads and those used in the transmission at high voltage of bulk electrical energy, as in the national grid.

Table 3.1 gives a list of the Regulations in the Electricity at Work Regulations that are particularly important to the issues surrounding in-service inspection and testing.

▼ **Table 3.1**
Regulations in the Electricity at Work Regulations relevant to in-service inspection and testing

Regulation 4	Systems, work activities and protective equipment
Regulation 5	Strength and capability of electrical equipment
Regulation 6	Adverse or hazardous environments
Regulation 7	Insulation, protection and placing of conductors
Regulation 8	Earthing or other suitable precautions
Regulation 10	Connections
Regulation 12	Means of cutting off the supply and for isolation
Regulation 13	Precautions for work on equipment made dead
Regulation 14	Work on or near live conductors
Regulation 15	Working space, access and lighting
Regulation 16	Persons to be competent to prevent danger and injury

Further information is given in Appendix III and in the HSE publication HSR 25.

3.1.5   Workplace (Health, Safety and Welfare) Regulations 1992

These regulations require that every employer shall ensure that the workplace equipment, devices and systems are maintained. This includes keeping the equipment devices and systems in an efficient state, in efficient working order, and in good repair. Where appropriate, the equipment, devices and systems shall be subject to a suitable system of maintenance.

The scope of the Workplace (Health, Safety and Welfare) Regulations is somewhat different to the Electricity at Work Regulations. The Electricity at Work Regulations are basically concerned with ensuring that an electrical installation is in a safe condition and that work performed on an electrical installation is done in a safe manner. They do not deal with the consequences of mal-operation of the electrical system. However, the Workplace Regulations are concerned with the consequences of equipment and system failures. For example, although a malfunctioning emergency lighting system may not in itself be an electrical hazard, there is a potential hazard if there is no emergency lighting. These regulations impose maintenance regimes upon such systems as emergency lighting, fire alarms, powered doors, escalators and moving walkways that have electrical power supplies. The regulations are not limited to electrical systems but also include equipment such as fencing, equipment used for window cleaning, devices to limit the opening of windows etc. The approved code of practice to the Workplace Regulations states that the maintenance of work electrical equipment and electrical systems is also addressed in other regulations. Electrical systems are clearly well addressed in the Electricity at Work Regulations and the maintenance of work equipment in the Provision and Use of Work Equipment Regulations 1998.

3.2   Scope of the legislation

The Health and Safety at Work etc. Act 1974, the Provision and Use of Work Equipment Regulations 1998 and the Electricity at Work Regulations 1989 apply to all electrical equipment used in, or associated with, places of work. The scope extends from distribution systems, be they 400 kV or those in buildings, down to the smallest piece of electrical equipment such as a hairdryer, a VDU, a telephone or even in some situations battery-operated equipment.

3.3   Who is responsible?

Everyone at work has responsibilities including, in certain circumstances, trainees. However, because of the all-embracing responsibilities of all persons this does not minimize the duties of particular persons. Regulation 3 of the Electricity at Work Regulations recognizes a responsibility (control) that employers and many employees have for electrical systems.

It shall be the duty of every employer and self-employed person to comply with the provisions of these Regulations in so far as they relate to matters that are within his/her control.

It shall be the duty of every employee while at work to

1 cooperate with his/her employer so far as is necessary to enable any duty placed on that employer by the provisions of these Regulations to be complied with, and
2 comply with the provisions of these regulations in so far as they relate to matters that are within his/her control.

The Provision and Use of Work Equipment Regulations 1998 require every employer to ensure that equipment is suitable for the use for which it is provided (Reg 4(1)) and only used for work for which it is suitable (Reg 4(3)). They require every employer to ensure equipment is maintained in good order (Reg 5) and inspected as necessary to ensure it is maintained in a safe condition (Reg 6).

This Code of Practice considers normal business premises such as shops, offices and workplaces and restricts advice to non-specialist installations and equipment that are commonly encountered.

3.4   Maintenance

Regulation 4(2) of the Electricity at Work Regulations 1989 states:

> As may be necessary to prevent danger, all systems shall be maintained so as to prevent, so far as is reasonably practicable, such danger.

Regulation 5 of the Provision and Use of Work Equipment Regulations 1998 states:

> Every employer shall ensure that work equipment is maintained in an efficient state, in efficient working order and in good repair.

The Approved Code of Practice & Guidance document to the Provision and Use of Work Equipment Regulations 1998 (L22) states that 'efficient' relates to how the condition of the equipment might affect health and safety; it is not concerned with productivity.

The Provision and Use of Work Equipment Regulations 1998 include a specific requirement that, where the safety of work equipment depends on installation conditions, and where conditions of work are liable to lead to deterioration, the equipment shall be inspected (Reg 6).

Fixed electrical installation

The safety and proper functioning of many portable appliances and items of equipment depend on the integrity of the fixed installation.

A system for the inspection and testing of the fixed installation as well as for the inspection and testing of portable appliances and equipment has to be established. The fixed installation will normally be subject to an initial inspection upon completion and a periodic inspection in accordance with the requirements of BS 7671. The initial inspection and subsequent periodic inspection and testing of the fixed installation will establish whether the installation is safe and fit for continued use. The frequency of the periodic inspection of the fixed installation depends on factors such as the type of premises and the use the installation receives. The frequency with which periodic inspection should be performed is given in the IEE's Guidance Note 3: *Inspection and Testing*. GN3 also provides information on the necessary competence of those carrying out the tests. A periodic inspection report identifying deficiencies and including inspection and test results should be issued to the responsible person.

Requirements placed by BS 7671 for the fixed installation, which have bearing on the in-service inspection and testing of electrical equipment, include the provision of an adequate means of earthing, RCD protection, numbers of socket-outlets and the condition of accessories.

4.1 Means of earthing

Metal parts of an appliance may become electrically charged if there is a fault. The purpose of earthing is to minimize the risk of electric shock should anyone touch those metal parts when a fault is present. This is achieved by providing a path for fault current to flow safely to Earth, via the circuit protective conductors in the fixed installation, which also causes the protective device to disconnect thereby removing the danger.

4.2 RCD protection

Portable equipment used outdoors is required to have RCD protection (Regulation 471-16-01 of BS 7671 refers) and the RCD used to provide this protection is required to have a residual operating current not exceeding 30 mA. The socket-outlet used to supply the portable equipment being used outdoors should have RCD protection and this is normally provided by an RCD installed in the consumer unit or distribution board. Alternatively, the socket-outlet itself can be a type that incorporates RCD protection. If there is any doubt, a plug-in portable RCD device/RCD adaptor should be employed (Figure 4.1).

▼ **Figure 4.1**
A plug-in RCD

4.3   Sufficient socket-outlets should be provided

Where portable equipment is likely to be used, provision should be made so that the equipment can be supplied from an adjacent and conveniently accessible socket-outlet, taking account of the length of the flexible cord normally fitted to a portable appliance or luminaire (Regulation 553-01-07 of BS 7671 refers). Extension leads should be avoided where possible and should be used only on a temporary basis. Where used, an extension lead is considered as a separate item of equipment and should be considered as a portable appliance.

4.4   Accessories should be undamaged

Cracked, broken, overheated and loose accessories, such as the socket-outlet illustrated in Figure 4.2, should be replaced.

▼ **Figure 4.2**
An overheated and cracked socket-outlet, which presents both a risk of fire due to overheating and a risk of electric shock due to possible contact with live parts

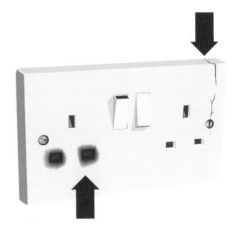

4.5   Specialized installations

Only suitable portable appliances and equipment are to be used in certain specialized fixed electrical installations, locations such as those with potentially explosive atmospheres, petrol filling stations and locations with combustible dusts. See Appendix I for information on relevant standards and for HSE Guidance Notes.

Types of electrical equipment

Refer also to the definitions given in Chapter 2 of this publication.

5.1 Portable appliances

A portable appliance is an appliance of not more than 18 kg in mass that is intended to be moved while in operation or an appliance that can easily be moved from one place to another, e.g. toaster, food mixer, kettle (Figure 5.1).

▼ **Figure 5.1**
A kettle is a portable appliance

5.2 Movable equipment (sometimes called transportable)

An item of movable equipment is equipment that is either

- ▶ 18 kg or less in mass and not fixed, e.g. an electric compressor (Figure 5.2), or
- ▶ equipment with wheels, castors or other means to facilitate movement by the operator as required to perform its intended use, e.g. air-conditioning unit.

▼ **Figure 5.2**
An electric compressor is an example of movable equipment

5.3   Hand-held appliances or equipment

A hand-held appliance or equipment is portable equipment intended to be held in the hand during normal use, e.g. angle grinder, power drill, hedge-cutter, soldering iron, iron, hairdryer as illustrated in Figures 5.3, 5.4 and 5.5.

▼ **Figure 5.3**
An iron is an example of a hand-held appliance

▼ **Figure 5.4**
An electric chainsaw is another example of a hand-held appliance

▼ **Figure 5.5**
An electric drill or breaker is a hand-held appliance

Electric shock can kill or seriously injure and is one of the hazards that electrical safety legislation is intended to protect against. Many serious shocks occur when the current flow is from hand to hand as the route the current follows is through or near the heart. Hand-held appliances present a particular danger as the appliance is gripped in one hand and it is quite possible that the other hand could be in contact with earthed metal. Being gripped, it becomes almost impossible to let go of the appliance under shock conditions. If the person using the appliance is sweating, such as a construction worker using an angle grinder, the contact resistance is significantly lowered and a larger current would flow. In addition, hand-held equipment can be the most prone to suffer misuse. Equipment used outdoors may be used under wet conditions or when the person has wet footwear thereby once again reducing contact resistance.

5.4 Stationary equipment or appliances

An item of stationary equipment or a stationary appliance is equipment that has a mass exceeding 18 kg and is not provided with a carrying handle, e.g. refrigerator, washing machine (Figure 5.6) or dishwasher.

▼ **Figure 5.6**
A washing machine is a stationary appliance

5.5 Fixed equipment or appliances

An item of fixed equipment or a fixed appliance is equipment that is fastened to a support or otherwise secured in a specified location, e.g. central heating boiler, hand dryer, fixed air-conditioning unit, bathroom heater, electric towel rail, immersion heater or luminaire (see Figure 5.7).

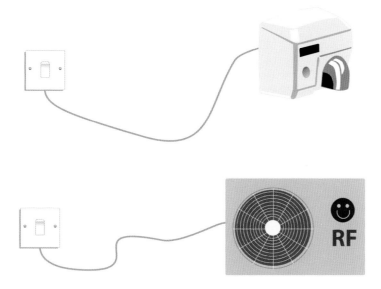

▼ **Figure 5.7**
Hand dryers and air-conditioning units are examples of fixed equipment

5.6 Appliances or equipment for building-in

An appliance or equipment for building-in is equipment intended to be installed in a prepared recess such as a cupboard or similar. In general, equipment for building-in does not have an enclosure on all sides, e.g. a built-in electric cooker.

5.7 Information technology equipment

Information technology (IT) equipment includes electrical business equipment such as computers and mains powered telecommunications equipment, and other equipment for general business use, such as mail processing machines, electric plotters, trimmers, VDUs, data terminal equipment, typewriters, telephones, printers, photocopiers and power packs.

5.8   Extension leads and RCD extension leads

An extension lead is necessary where an item of equipment needs to be supplied but a convenient socket-outlet is not available. An RCD extension lead is an extension lead that includes an RCD.

5.9   Multiway adaptors and RCD adaptors

Multiway adapters and cube adaptors are used when there are not sufficient socket-outlets available.

RCD adaptors are used to provide protection for persons using portable equipment, particularly for persons using portable equipment outdoors.

The electrical tests   6

6.1   Testing throughout the life of equipment

Electrical equipment may be subjected to tests at four stages of its life:

1   manufacturer's type testing to an appropriate standard
2   manufacturer's production testing
3   in-service inspection and testing (covered by this Code of Practice)
4   testing after repair

6.2   Manufacturer's type testing

Test houses or the manufacturer carry out type testing to assess compliance with a standard (British or European). The tests are usually destructive, making that particular appliance unsuitable for sale or use.

6.3   Manufacturer's production testing

Manufacturers carry out production testing to ensure that appliances are in accordance with the appropriate standard.

New appliances uncontaminated by dust or lubricants are subjected to production tests. Equipment in new or as-new condition may also be subjected to production testing following refurbishment or repair. Further information on manufacturer's production testing is given in Appendix V.

Dielectric strength testing (also known as hi-pot or flash testing) should only be performed where it will not weaken insulation or create a hazard. If in doubt seek advice from the manufacturer of the appliance.

6.4   In-service inspection and testing

In-service inspection and testing as detailed in this Code of Practice are carried out on a routine basis to determine whether the particular item of equipment is in a safe condition for continued use. It is normally not necessary to test new items of equipment as the manufacturer has already tested them. In-service inspection and testing are not as onerous as production testing.

Three categories of in-service inspection and testing are referred to in this Code of Practice.

1   **User check:** performed by the user before plugging in and switching on the equipment and involves a visual inspection that includes the plug, the flex, the appliance and the suitability of the appliance for the environment and the job. Faults or suspected faults should be reported and logged, and faulty equipment should be removed from service. Normally, no record is required if no fault is found. Refer to Chapter 13 of this publication.

2 **Formal visual inspection:** performed by a competent person, often the test operative. The formal visual inspection consists of performing the user checks and, in addition, checks on the suitability of the equipment for the environment and checks on the switching of the equipment. The formal visual inspection includes removing the plug top and inspecting the wiring, connections, fuse and cord grip for rewireable plugs. Equipment failing the formal visual inspection should be removed from service, labelled and repaired. The formal visual inspection should be recorded. Refer to Chapter 14 of this publication.

3 **Combined inspection and test:** performed by the test operative and consist of
i   an inspection,
ii testing as necessary to ensure the equipment is in a safe condition (the testing may include earth continuity testing at a high or low test current and insulation testing or a protective conductor/touch current measurement), and
iii functional checks.

The combined inspection and test should be recorded. Refer to Chapter 15 of this publication.

6.5   Testing after repair

An item of electrical equipment that has been repaired should be inspected and tested in accordance with either the production tests covered in Section 6.3 or the in-service tests covered in Section 6.4. The decision should be based on the type of equipment, its condition and the nature of the repairs (Figure 6.1).

Guidance on testing after repair may be obtained from the manufacturer.

The inspection and testing should be recorded.

▼ **Figure 6.1**
An item of equipment, such as this electric breaker, that has been repaired should be inspected and tested in accordance with either the production tests covered in Section 6.3 or the in-service tests covered in Section 6.4

In-service inspection and testing | 7

7.1 Inspection

In-service inspection and testing of equipment are essential to ensure safety, and a regime of time and risk assessment based inspections and tests should be implemented. In-service inspections can often be carried out by the user of the equipment, and in some circumstances this may be all that is necessary. An example of circumstances where user inspections may be the only inspection required is in a low risk environment where Class II equipment is used.

Inspection should always precede testing.

A properly carried out inspection can identify many faults that will not necessarily be apparent from electrical tests, such as a cracked case, a loose connection, a damaged flex and evidence of overheating.

7.2 Categories of inspection and testing

Three categories of in-service inspection and testing are referred to in this Code of Practice.

1 **User checks:** faults are to be reported and logged and faulty equipment should be removed from service. No record is required if no fault is found. Refer to Chapter 13.
2 **Formal visual inspections:** inspections without tests, the results of which, satisfactory or unsatisfactory, are recorded. Refer to Chapter 14.
3 **Combined inspections and tests:** the results of which are recorded. Details of these inspections and tests are provided in Chapter 15.

7.3 Frequency of inspection and testing

The relevant requirement of the Electricity at Work Regulations 1989 is that equipment shall be maintained so as to prevent danger. Inspection and testing are means of determining whether maintenance is required. The frequency of inspection and testing will depend upon the likelihood of maintenance being required and the consequence of lack of maintenance. No rigid guidelines can be laid down, but factors influencing the decision include the following.

1 **The environment:** equipment installed in a benign environment, such as an office, will suffer less damage than equipment in an arduous environment, such as a construction site.
2 **The users:** if the users of equipment report damage as and when it becomes evident, hazards will be avoided. Conversely, if equipment is likely to receive unreported abuse, more frequent inspection and testing are required.

▼ **Table 7.1** Initial frequency of inspection and testing of equipment

	Equipment use	Type of equipment	User checks	Class I		Class II	
			Not recorded unless a fault is found	Formal visual inspection (Note 1) Recorded	Combined inspection and testing Recorded	Formal visual inspection (Note 1) Recorded	Combined inspection and testing Recorded
	a	b	c	d	e	f	g
1	Construction sites 110 V equipment	S	none	1 month	3 months	1 month	3 months
		IT	none	1 month	3 months	1 month	3 months
		M	weekly	1 month	3 months	1 month	3 months
		P (Note 2)	weekly	1 month	3 months	1 month	3 months
		H	weekly	1 month	3 months	1 month	3 months
2	Industrial including commercial kitchens	S	weekly	none	12 months	none	12 months
		IT	weekly	none	12 months	none	12 months
		M	before use	1 month	12 months	3 months	12 months
		P	before use	1 month	6 months	3 months	6 months
		H	before use	1 month	6 months	3 months	6 months
3	Equipment used by the public	S	Notes 3,4	monthly	12 months	3 months	12 months
		IT	Notes 3,4	monthly	12 months	3 months	12 months
		M	Notes 3,4	weekly	6 months	1 month	12 months
		P	Notes 3,4	weekly	6 months	1 month	12 months
		H	Notes 3,4	weekly	6 months	1 month	12 months
4	Schools (Note 4)	S	weekly	none	12 months	12 months	48 months
		IT	weekly	none	12 months	12 months	48 months
		M	weekly	4 months	12 months	4 months	48 months
		P	weekly	4 months	12 months	4 months	48 months
		H	before use	4 months	12 months	4 months	48 months
5	Hotels (Note 5)	S	none	24 months	48 months	24 months	None
		IT	none	24 months	48 months	24 months	None
		M	weekly	12 months	24 months	24 months	None
		P	weekly	12 months	24 months	24 months	None
		H	before use	6 months	12 months	6 months	None
6	Offices and shops	S	none	24 months	48 months	24 months	None
		IT	none	24 months	48 months	24 months	None
		M	weekly	12 months	24 months	24 months	None
		P	weekly	12 months	24 months	24 months	None
		H	before use	6 months	12 months	6 months	None

Refer to notes opposite

3 **The equipment construction:** the safety of a Class I appliance is dependent upon a connection with the earth of the fixed electrical installation. If the flexible cable is damaged the connection with earth can be lost.
   The safety of Class II equipment is not dependent upon the integrity of the electrical installation.
   If equipment is known to be Class II and is used in a low risk environment, such as an office, recorded testing (but not inspection) may be omitted – see Table 7.1.

4 **The equipment type:** An appliance that is hand-held is more likely to be damaged than a fixed appliance. If such an appliance is also Class I the risk of danger is increased, as safety is dependent upon the continuity of the protective conductor from the plug to the appliance.

Table 7.1 provides guidance on *initial* frequencies of inspection and testing. The frequency of inspection and testing depends upon the factors above, i.e. any circumstance that may affect the continuing safety of the equipment.

Notes to Table 7.1:
1 The formal visual inspection may form part of the combined inspection and tests when they coincide, and is to be recorded.
2 110 V earthed centre-tapped supply. 230 V portable or hand-held equipment is required to be supplied via a 30 mA RCD and inspections and tests carried out more frequently.
3 For some equipment such as children's rides a daily check may be necessary.
4 By supervisor/teacher/member of staff.
5 Equipment provided in hotel rooms is equipment used by the public. Equipment in hotels (row (5) above) is considered to be equipment in hotels used by the hotel staff.
▶ It is normally not necessary to test new items of equipment as the manufacturer has already tested them.
▶ The information on suggested initial frequencies given above is more detailed and specific than HSE guidance, but is not considered to be inconsistent with it.

    S Stationary equipment
    IT Information technology equipment
    M Movable equipment
    P Portable equipment
    H Hand-held equipment

The frequency of any recurring damage should be noted and corrective action taken. Corrective action to be considered should include

- ▶ replacement of the equipment with a more rugged type,
- ▶ training for the people using the equipment, and
- ▶ increasing the frequency of inspection and testing.

The most important check that can be carried out on a piece of equipment is the visual inspection. The visual inspection can identify many defects, particularly in the case of portable appliances or hand-held tools where defects in the plug, the cable or the casing could occur.

If the user cannot routinely disconnect the equipment to facilitate a user inspection, this should be taken into account when determining the frequency of recorded inspection.

Where premises have mixed use, the most appropriate frequency of inspection and testing will need to be adopted for each location or use.

7.4　Review of frequency of inspection and testing

The intervals between checks, formal inspections and tests should be kept under review, particularly until patterns of failure or damage, if any, are determined.

Particularly close attention should be paid to initial checks, formal inspections and tests to see if there is a need to reduce the intervals or change the equipment or its use.

After the first few inspections and tests consideration should be given to increasing the intervals or reducing them.

Procedures for in-service inspection and testing

8

8.1 The basic requirement

Regulation 4(2) of the Electricity at Work Regulations 1989 requires that:

> As may be necessary to prevent danger, all systems shall be maintained so as to prevent, so far as is reasonably practicable, such danger.

The legal duty is to ensure that if danger would result, electrical equipment is maintained to prevent such danger. There is no specific requirement for inspection or testing. However, inspection is implicit when carrying out maintenance, to confirm the safe condition or identify defects. Furthermore, some faults can only be identified by testing. A maintenance regime comprising user checks, formal visual inspection and, where appropriate, testing will help prevent accidents. Record keeping will demonstrate that a maintenance regime exists.

The requirements of the Electricity at Work Regulations are not necessarily met simply by carrying out, inspections and tests on equipment. The aim is to ensure that equipment is maintained in a safe condition. Faulty equipment is required to be maintained or replaced. Records should be kept.

Organisations whose equipment is being inspected and tested will need to assist in identifying the tests to be carried out, including any variances from the guidance in this document. Manufacturers' recommendations and past test results must be requested by the test operative from the responsible person. This is particularly appropriate to business and telecommunications equipment, where it may be necessary to determine the appropriate tests with the supplier of the equipment. Some tests, particularly insulation resistance tests, may damage such equipment. If there is doubt, minimum tests, such as the 'soft tests' in Chapter 15, should be carried out, and the customer advised that, after discussion with the supplier and downloading of data, it may be necessary to carry out further tests.

8.2 Test and repair equipment

Test equipment will be needed that will perform, as a minimum, an earth continuity test and an insulation resistance test to permit testing, as necessary, of all the types of electrical equipment likely to be encountered so that necessary maintenance and repair are identified. Refer to Chapter 10 for further details on test instruments.

Following inspection and testing, certain repairs will be identified, and the test operative will need suitable hand tools and spare parts such as plugs, fuses and replacement cable.

8.3   Documentation

Although there is no requirement in the Electricity at Work Regulations 1989 to keep records of equipment and of inspections and tests, the HSE Memorandum of Guidance on these regulations advises that records of maintenance including tests should be kept throughout the working life of equipment. These records are a useful management tool for reviewing the frequency of inspection and testing, and without such records dutyholders cannot be certain that the inspection and testing have actually been carried out.

The following records (see Appendix VI) should be established and maintained.

1   A register of all equipment (Form VI.1) – new equipment, as it is purchased, should be added to this list.
2   A record of formal and combined inspections and tests (Form VI.2).
3   A repair register (Form VI.4).
4   A register of all faulty equipment (Form VI.5).
5   All equipment formally inspected and tested should be labelled as per Section 8.4 (Form VI.3).

These records may be retained on paper or in electronic memory providing reasonable precautions are taken with respect to security. Previous test results should be made available to subsequent test operatives.

The following records should be maintained by the organisation carrying out the inspection and testing, and copies given to the responsible person.

1   Copy of the formal visual inspection and combined inspection and test results.
2   Register of all equipment repaired.

These may be held as paper or electronic records.

8.4   Labelling

All equipment that requires routine inspection and/or testing should be clearly identifiable. This is usually achieved by labelling of the equipment. The information provided should consist of an identification code to enable the equipment to be uniquely identifiable even if several similar items exist within the same premises. An indication of the current safety status of the equipment should also be included (e.g. whether the item has PASSed or FAILed the appropriate safety inspection/test). The date on which re-testing is due or the last test date and re-test period should also be stated.

The provision of the above information will not only make it easier to locate the equipment at the time of re-test but will enable non-technical users to become aware of any equipment that is due for re-test, or that should not be used because testing is overdue.

Additional information may also be included such as the company name or logo. Labels may either be pre-printed and filled in by hand or be machine-readable (e.g. bar coded). The latter is particularly suitable for the identification code. Many test operatives have equipment that can read bar coded labels in order to set up the instrument to conduct only the tests appropriate for the equipment. Pass/fail information and date should be in text format so that they are readable by others.

Labels may take many forms but should be such that they can be reliably applied to a variety of surfaces. They should be durable and capable of surviving the period between re-tests without undue degradation. In industrial environments the demands on the label are high since it may be subject to contact with oils, solvents, moisture and abrasion. The label should be fixed in a prominent position where it can be clearly seen.

In order to keep proper records, items such as extension leads, which may not have serial numbers, will have to be identified with a unique reference number or code fixed to, or marked on, the equipment.

8.5   Damaged or faulty equipment

If equipment is found to be damaged or faulty on inspection or test, it must be removed from service and then an assessment should be made by a responsible person as to the suitability of the equipment for the use in that particular location. More frequent inspections and tests will not prevent damage to equipment arising if it is unsuitable for the use, location or environment. If it is unsuitable it should be replaced by more suitable equipment.

8.6   User responsibilities

Users have a duty to do what is reasonably practicable to ensure that equipment is safe. Persons and users of equipment should be advised that it is their legal responsibility to comply with the Health and Safety at Work etc. Act and the Electricity at Work Regulations by assisting in the maintenance of equipment.

Users should be aware that equipment will be regularly inspected, tested and labelled.

Users should be aware that

▶ faulty equipment should NOT be used,
▶ faulty equipment should be labelled as such, reported and withdrawn from service, and
▶ they should seek to identify any reason why equipment has become faulty, such as abuse or unsuitability and take appropriate action.

8.7   Provision of test results

Previous inspection and test results should be provided to operatives carrying out in-service inspection and testing to assist in identifying any deterioration that may have occurred.

Training 9

9.1 The system

Regulation 4(2) of the Electricity at Work Regulations requires that electrical equipment is to be maintained so as to prevent, so far as is reasonably practicable, danger. An organisation will normally meet this statutory requirement by setting up a system of inspection and testing and repair of electrical equipment involving user checks, formal visual inspections, combined inspections and tests, and repair and replacement to ensure that equipment does not present danger. The system employed places responsibilities on:

- ▶ The user, the person about to use the electrical equipment. The user should check the equipment before use.
- ▶ The administrator, normally a manager or supervisor, who is responsible for setting up and maintaining the system of inspection, testing, repair, replacement and training.
- ▶ The inspector, often the test operative, who performs the formal visual inspection on a routine basis.
- ▶ The test operative who inspects and tests electrical equipment on a routine basis.
- ▶ The person repairing faulty electrical equipment.
- ▶ Other dutyholders such as company directors, managers and building service managers.

Regulation 16 of the Electricity at Work Regulations requires that persons are competent to perform their duties. For example, the persons undertaking the inspection and, where appropriate, testing of electrical equipment and appliances should be competent to do so having due regard to their own safety and that of others. The above division of tasks and responsibilities for specifically trained persons does not presume that one person may not carry out two or more of the required functions. It is, however, stressed that a person should be trained in each of the areas and competent to undertake the work and interpret the results as appropriate.

9.2 The user

Users may need to be trained to look for defects that can occur in electrical equipment. Users should visually inspect electrical equipment for defects before it is switched on and used.

Users should be aware that

- ▶ faulty equipment or equipment suspected of being faulty should not be used, and
- ▶ faulty equipment should be labelled, reported and removed from use without delay.

Further information and drawings of common deficiencies are given in Chapter 13 of this publication.

9.3   The administrator or manager

Administrators or managers of premises are required to know their legal responsibilities as laid down in the Electricity at Work Regulations 1989. They should understand and apply the legislation and assess the risks in respect of electrical equipment and appliances within their charge or which they are contracted to inspect, test, and repair or replace. Administrators have a legal responsibility to ensure that the electrical equipment in their charge is safe. Training may therefore be needed to understand Sections 1 to 8 of this Code of Practice.

Administrators may also require training so that they can maintain the records of inspections, tests, and repairs or replacement of appliances and equipment and manage the re-inspection and re-testing at appropriate intervals as specified. Administrators are required to interpret the recorded results and take appropriate actions regarding equipment or to provide a report to a more senior person within the organisation. Competence to interpret recorded results is achieved by appropriate training and experience.

9.4   The inspector

The inspector, often the test operative, should be competent to perform the formal visual inspection and record the results. Training may be required. The inspector should be prepared to fill in and sign the necessary reports and state that an item of equipment is safe or unsafe for continued use.

The formal visual inspection includes checking the cable and the plug, including its internal wiring if practicable. Further information is given in Chapter 14 of this Code of Practice.

9.5   The test operative

The test operative should be competent to inspect and test an item of equipment and, based on the results and upon the conditions in which it is being used, state that the equipment is safe or otherwise for continued use. Training and experience will both be necessary.

9.5.1   Training

The test operative should have training that includes the identification of equipment and appliance types to determine the test procedures and frequency of inspection and testing. Test operatives should be trained so as to be familiar with the test instruments used and in particular their limitations and restrictions so as to achieve repeatable results without damaging the equipment. The test operative should be able to fill in records and sign them to take responsibility for the work.

The test operative should be trained to have an understanding of how electrical, mechanical or thermal damage can occur to electrical equipment, flexes and plugs and connections.

9.5.2 Experience

The test operative should have sufficient experience and technical knowledge to perform the inspection and testing without putting him/herself or others at risk. Technical knowledge or experience may consist of adequate knowledge of electricity and adequate experience of electrical work. The test operative should have an adequate understanding of the equipment to be worked on and practical experience of that system. The operative should be aware of the hazards that may arise during the work and the precautions that need to be taken, and should be able to recognize at all times whether it is safe for work to continue. Alternatively, if the operative is learning the job, that person should be adequately supervised.

The test operative should be prepared to declare that an item of equipment is safe for continued use. No-one can guarantee that an item of equipment will remain safe for the next year; this is not possible as the item may be dropped or otherwise physically damaged the following day. The test operative can only say that on a particular day a particular item is safe for use and should be safe for continued use under the particular conditions. The test operative should be prepared to advise if a certain item of equipment is unsuitable for the particular location or external influences and should be replaced with a more rugged item. The test operative should also be prepared to advise on a cost-effective maintenance regime including the periods between formal visual inspections and combined inspections and tests.

9.6 The person repairing faulty equipment

The person repairing faulty equipment should be able safely and effectively to repair the faulty item of equipment without putting themselves or others at risk. Once again, suitable training and experience will be needed. Some repairs may be straightforward, such as changing a fuse, a plug or a cable; others may be more complicated requiring knowledge of the particular item of faulty equipment. Some items may have to be returned to the manufacturer for repair. In all cases, records should be kept and the person repairing the equipment will have to take responsibility for their work by signing the appropriate certificate. Records of maintenance, including test results, preferably kept throughout the working life of an electrical system will enable the condition of the equipment and the effectiveness of maintenance policies to be monitored. Without effective monitoring dutyholders cannot be certain that the requirement for maintenance has been complied with.

Test instruments 10

10.1 Safety of test equipment

Test equipment complete with suitable probes and leads is required to carry out the required tests.

10.1.1 Test instruments

All test instruments should be safe. The current safety standard is BS EN 61010: *Safety requirements for electrical equipment for measurement, control, and laboratory use*. All new equipment should comply with this or an equivalent standard. Equipment pre-dating this standard is not necesssarily unsafe.

The operating instructions and advice given by the manufacturer should be read and followed when using test instruments.

Generally, portable appliance test instruments offer the most convenient means of providing the required test facilities but this does not preclude the use of suitable individual general test instruments.

10.1.2 Test probes and leads

The product standard for test probes and leads is BS EN 61010-031.

10.1.3 Test probes and leads for use in conjunction with a voltmeter, multimeter, electrician's test lamp or voltage indicator

In addition to the requirements of BS EN 61010-031: *Safety requirements for electrical equipment for measurement, control, and laboratory use*, test probes and leads should comply with the recommendations given in HSE Guidance Note GS38.

10.2 Portable appliance test instruments

Most portable appliance test instruments (PATs) provide the following facilities:

▶ measurement of earth continuity with one or more pre-set test currents up to a maximum value of the order of 26 A
▶ measurement of insulation resistance normally using a test voltage of 500 V d.c.

Portable appliance test instruments (Figure 10.1) may offer additional test facilities such as:

▶ measurement of earth continuity using a low value of current in the range 20 mA to 200 mA, typically 100 mA, known as a 'soft test'

▶ insulation resistance assessment by the protective conductor/touch current* method, known as a 'soft test'

▶ a load test

▶ dielectric strength or flash testing

▶ substitute/alternative leakage measurement

▶ insulation resistance at 250 V d.c.

* Refer to the definitions in Chapter 2 for 'Protective conductor current' and 'Touch current'.

▼ **Figure 10.1**
Portable Appliance Test Instrument [Illustration courtesy of Martindale Electric]

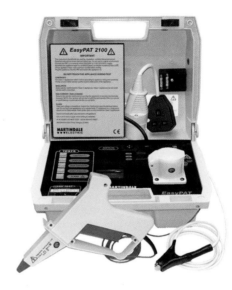

The substitute/alternative leakage measurement is made between both the live and neutral conductors connected together and protective earth. The test voltage is a.c. with a frequency of 50 Hz, which means the leakage paths will be similar to when the equipment is in operation.

The substitute/alternative leakage measurement is not a replacement for protective conductor or touch current measurement but can be useful in some situations if the limitations are clearly understood

The test method is similar to insulation testing. The measurement is made between both phase conductors and protective earth but the test voltage is 50 Hz a.c., which means that, unlike insulation resistance testing, the impedance of leakage paths will be the same as when the equipment is in operation. Similarly, because the magnitude of the test voltage is not greater than the nominal supply voltage, measurements are not affected by voltage limiting devices.

The substitute/alternative leakage test has some limitations because

▶ electronic switches will not be on, and

▶ relays or other active circuitry that may effect measurements may not be activated.

In the case of electronic switches the appliance will not be tested beyond the switch.

The test should be performed in addition to an insulation test.

10.2.1 Three-phase equipment

If the equipment is fitted with a plug complying with BS EN 60309: *Plugs, socket-outlets and couplers for industrial purposes* a three-phase/single-phase conversion lead can be made up and employed to enable a standard PAT to be used. Such special test leads must be kept under the control of the test operative. Alternatively, an insulation/continuity test instrument can be used (see Sections 15.4 and 15.5), testing at the equipment terminals once the supply has been isolated and proved dead.

10.3 Low resistance ohmmeters (for earth continuity testing)

Earth continuity testing may in certain circumstances (see Section 15.4 of Part 2) be carried out by a low resistance ohmmeter (Figure 10.2). The ohmmeter may be either a specialized low resistance ohmmeter, or the continuity range of a combined insulation and continuity test instrument. The test current may be a.c. or d.c., but should be derived from a source with an open-circuit voltage of not less than 100 mV and no greater than 24 V. The test current should be within the range 20 to 200 mA nominal. Instruments conforming to BS EN 61557-4: *Electrical safety in low voltage distribution systems. Equipment for testing, measuring and monitoring of protective measures* are adequate.

▼ **Figure 10.2**
Low resistance ohmmeter [Illustration courtesy of Seaward Electronics Ltd]

The resolution at low scale values should be 0.01 Ω. A basic instrument accuracy in accordance with BS EN 61557-4 is adequate. Inaccuracies can be caused by:

► **Contact resistance,** which cannot be eliminated with a normal two terminal test instrument and can introduce an error of 0.1 Ω or more. Test probes should be kept in good condition and applied firmly during measurement.

► **Test lead resistance,** which can be eliminated by measuring the resistance of the leads prior to a test, and subtracting the resistance from the final value. Test lead resistance may alternatively be 'nulled' prior to the test where the test instrument includes this facility. Fuses in test leads will add a certain amount of resistance.

► **Crocodile clips,** which can cause inaccuracies. Faulty, loose, corroded or damaged clips should be replaced.

► **Joints,** which will introduce resistance. Joints should be avoided wherever possible.

▶ **Interference from an a.c. source,** which can only be eliminated by locating the source, such as a nearby switched mode power supply or transformer, and switching it off.

▶ **Thermocouple effects in mixed metal systems,** which can be eliminated by reversing the test probes and averaging the resistance readings taken in each direction.

10.4 Insulation resistance ohmmeters (applied voltage method)

The recommended test voltage is 500 V d.c. The instrument used should be capable of maintaining the test voltage when applied to the insulation of the equipment under test. Instruments conforming to BS EN 61557-2 meet these recommendations.

If the insulation test gives a zero or very low value, it is an indication that the equipment being tested has a fault and the fault should be rectified before any further testing.

Some types of equipment may have filter networks or transient suppression devices. Such equipment may give values of insulation resistance below normally accepted levels due to the presence of these components by design. On such equipment, a protective conductor current test (Class I devices, also sometimes referred to as a leakage test) and a touch current test (Class I and Class II devices) are recommended in addition to the insulation test.

Additional tests may be needed between phases and between conductors.

Note that some equipment may not be suitable for insulation testing at 500 V, particularly older equipment not complying with EN 60950. In such cases the manufacturer of the equipment should be consulted before proceeding with the 500 V insulation test and also before proceeding with alternative test options such as a reduced insulation test at 250 V. If powered tests are to be conducted, such as the protective conductor or touch current test, it should be confirmed with the manufacturer that proceeding with such tests is safe.

10.5 Dielectric strength testing

Dielectric strength testing (also known as flash or hi-pot testing) is not normally carried out during in-service testing. Dielectric strength testing is carried out by the manufacturer on a complete appliance after assembly.

Dielectric strength testing should only be performed where it will not weaken insulation or create a hazard. If in doubt seek advice from the manufacturer of the appliance.

10.6 Instrument accuracy

The accuracy of a test instrument should be verified and recorded annually or in accordance with the manufacturer's instructions. Test instruments are often calibrated on an annual basis by sending the instrument to a calibration house which calibrates the instrument and provides a certificate stating the date of calibration, the time period for which the certificate would be valid along with the findings of the assessment in the form of a table of results. The certificate verifies that the instrument is within calibration parameters at that time only. The certificate does not guarantee that the instrument is still fit-for-purpose at any time after that. In addition, it is recommended that the test instrument is checked at regular intervals.

One method of assessing the on-going accuracy of test instruments is to maintain records, over time, of measurements taken from designated reference circuits or items of equipment such as a resistance box or voltage source.

In each instance, the designated circuit or item of equipment should be used for every subsequent assessment.

Before such a system is implemented, the accuracy of each test instrument should be confirmed and this can only be carried out by a formal calibration house. Test leads should be assessed at the time of calibration.

To avoid ambiguity, the relevant testing points should labelled allowing other operatives, who may not usually be charged with the task of test instrument assessment, to follow the system. Should the results waver by ± 5 per cent, the instrument should be recalibrated.

Many test instrument manufacturers produce proprietary 'checkboxes' that incorporate many testing functions.

A record sheet should be kept; see Form VI.6 in Appendix VI.

Part 2
Inspection and Testing (including user checks)

Equipment constructions

Table 11.1 lists the different types of equipment construction that are likely to be encountered.

◄ **Table 11.1**

Section	Type	Figure
11.1	**Class I**	
11.1.1	Basic insulation and earthed metal	11.1
11.1.2	Air used as a basic insulation medium	11.2
11.1.3	Unearthed metal separated from live parts by basic and supplementary insulation	11.3
11.1.4	Unearthed metal separated from live parts by basic insulation and earthed metal	11.4
11.2	**Class II**	
11.2.1	Equipment with a substantial enclosure of insulating material comprising basic and supplementary insulation	11.5
11.2.2	Equipment with a substantial enclosure of reinforced insulating material	11.6
11.2.3	Equipment with a substantial enclosure of insulating material – the insulation construction includes air	11.7
11.2.4	Equipment with unearthed metal in the enclosure, separated from live parts by basic and supplementary insulation	11.8
11.2.5	Equipment with unearthed metal separated from live parts by reinforced insulation	11.9
11.2.6	Equipment with unearthed metal separated from live parts by basic and supplementary insulation including air gaps	11.10
11.2.7	Metal-encased Class II equipment	11.11
11.3	**Class III**	
11.4	**Class 0 and Class 0I**	

There are a number of basic equipment constructions that are referred to in all standards for electrical equipment and in this Code of Practice. They are important because they determine how the user is protected against electric shock and describe tests appropriate to apply when assessing safety. Appliances are often not what they appear, and so a number of typical constructions are shown.

11.1 Class I

Class I equipment includes appliances and tools, and for such equipment, protection against electric shock is provided by both

- ▶ the provision of basic insulation, and
- ▶ connecting metal parts to the protective conductor in the connecting cable and plug and hence via the socket-outlet to the fixed installation wiring and the means of earthing.

The metal parts could assume a hazardous voltage if the basic insulation should fail, hence the requirement that the metal parts are earthed via the protective conductor.

Class I equipment may have parts with double insulation or reinforced insulation or parts operating at extra-low voltage.

Class I equipment relies for its safety upon a satisfactory means of earthing for the fixed installation and an adequate connection to it, normally via the flexible cable connecting the equipment, the plug and socket-outlet, and the circuit protective conductors of the fixed installation. Where Class I equipment is intended to be used with a flexible cable, the cable is required to include a protective conductor.

11.1.1 Class I typical construction showing basic insulation and earthed metal

▼ **Figure 11.1**
Class I equipment
showing basic insulation
and earthed metal

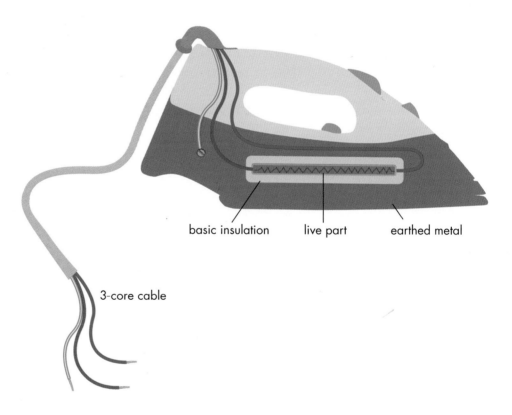

basic insulation   live part   earthed metal

3-core cable

11.1.2 Class I construction showing the use of air as a basic insulation medium

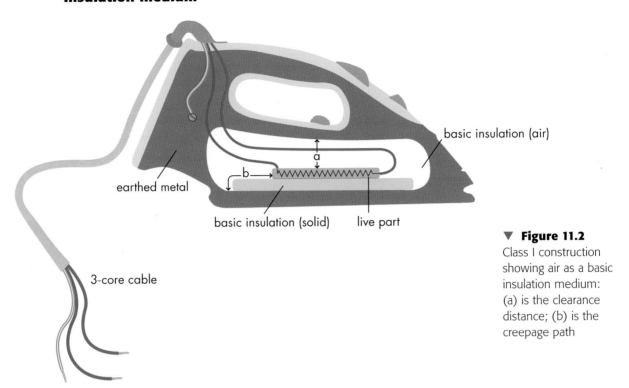

basic insulation (air)

earthed metal

basic insulation (solid) live part

3-core cable

▼ **Figure 11.2**
Class I construction showing air as a basic insulation medium: (a) is the clearance distance; (b) is the creepage path

11.1.3 Class I construction incorporating unearthed metal separated from live parts by basic and supplementary insulation

Unearthed metal may be encountered in Class I appliances as shown in Figures 11.3 and 11.4. In Figure 11.3 the unearthed metal will not be connected to the protective conductor. In Figure 11.4 the unearthed metal may be in casual or fortuitous contact with the earthed metal. A continuity test made to this 'unearthed' metal may give misleading test results.

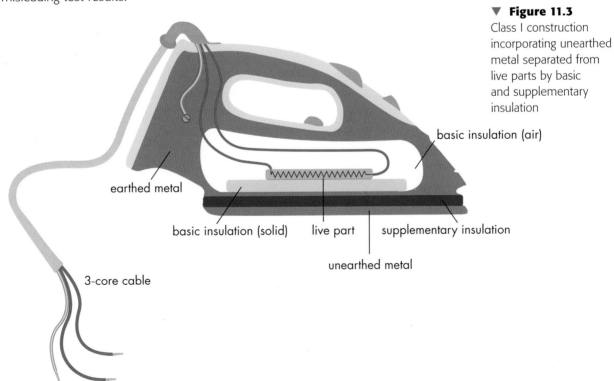

▼ **Figure 11.3**
Class I construction incorporating unearthed metal separated from live parts by basic and supplementary insulation

basic insulation (air)

earthed metal

basic insulation (solid) live part supplementary insulation

unearthed metal

3-core cable

11.1.4 Class I construction incorporating unearthed metal separated from live parts by basic insulation and earthed metal

▼ **Figure 11.4**
Class I construction incorporating unearthed metal separated from live parts by basic insulation and earthed metal

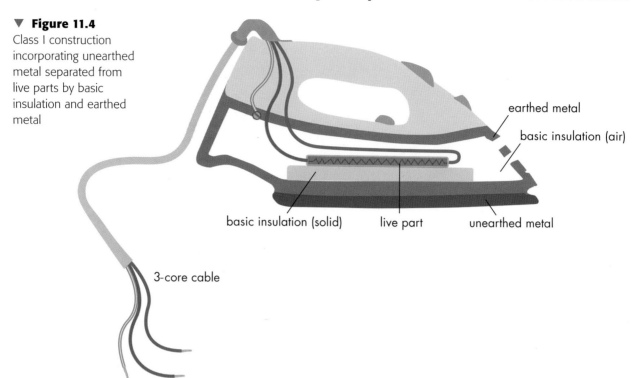

earthed metal

basic insulation (air)

basic insulation (solid)   live part   unearthed metal

3-core cable

11.2   Class II

Class II equipment is equipment in which protection against electric shock is provided by

▶ basic insulation and an additional safety precaution such as supplementary insulation, or
▶ reinforced insulation.

Protection against electric shock does not rely on protective earthing. The equipment does not rely on the circuit protective conductors and the means of earthing provided in the fixed installation.

Class II equipment may be insulation-encased or metal-encased.

▶ **Insulation-encased Class II.** Equipment having a durable and substantially continuous electrical enclosure of insulating material that envelops all conductive parts with the exception of small parts such as name plates, drill chucks, screws and rivets, which are insulated from live parts by insulation at least equivalent to reinforced insulation. Insulation-encased Class II equipment is illustrated in Figures 11.5, 11.6 and 11.7. The enclosure of an insulation-encased Class II appliance may form part or the whole of the supplementary insulation or reinforced insulation.

▶ **Metal-encased Class II.** Class II equipment may have a substantially continuous metal enclosure with double or reinforced insulation used throughout (Figure 11.11).

11.2.1 Class II equipment with a substantial enclosure of insulating material comprising basic and supplementary insulation

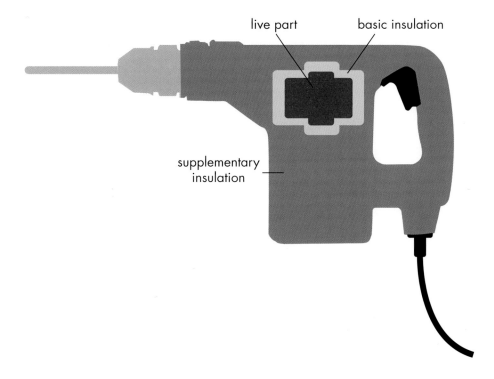

live part   basic insulation

supplementary insulation

▼ **Figure 11.5**
Class II equipment with a substantial enclosure of insulating material comprising basic and supplementary insulation (Insulation-encased Class II)

11.2.2 Class II equipment with a substantial enclosure of reinforced insulating material

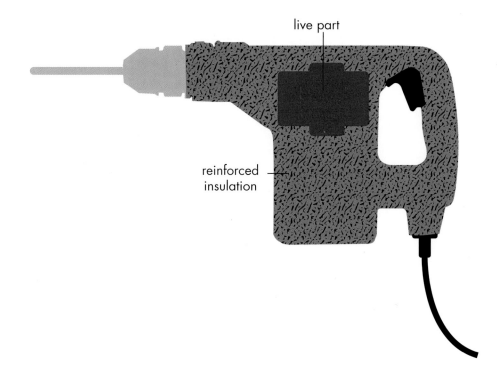

live part

reinforced insulation

▼ **Figure 11.6**
Class II equipment with a substantial enclosure of reinforced insulating material (Insulation-encased Class II)

11.2.3 Class II equipment with a substantial enclosure of insulating material – the insulation construction includes air

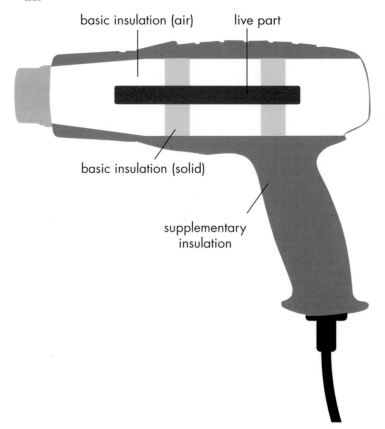

basic insulation (air)   live part

basic insulation (solid)

supplementary insulation

▼ **Figure 11.7**
Class II equipment
with a substantial
enclosure of insulating
material – the insulation
construction includes
air (Insulation-encased
Class II)

11.2.4 Class II equipment with unearthed metal in the enclosure, separated from live parts by basic and supplementary insulation

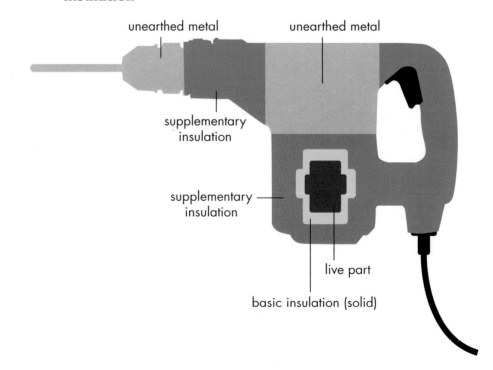

unearthed metal   unearthed metal

supplementary insulation

supplementary insulation

live part

basic insulation (solid)

▼ **Figure 11.8**
Class II equipment with
unearthed metal in the
enclosure separated
from live parts by basic
and supplementary
insulation

11.2.5 Class II equipment with unearthed metal separated from live parts by reinforced insulation

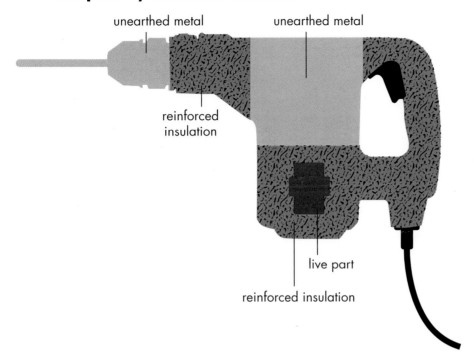

unearthed metal

unearthed metal

reinforced insulation

reinforced insulation

live part

▼ **Figure 11.9**
Class II equipment with unearthed metal separated from live parts by reinforced insulation

11.2.6 Class II equipment with unearthed metal separated from live parts by basic and supplementary insulation including air gaps

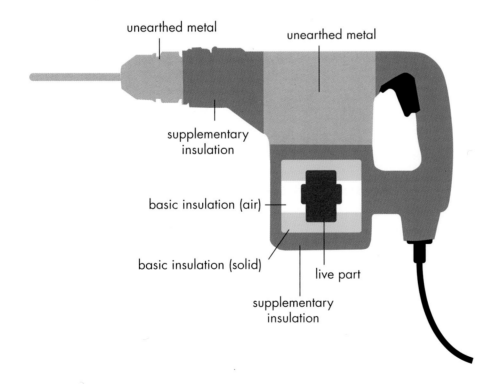

unearthed metal

unearthed metal

supplementary insulation

basic insulation (air)

basic insulation (solid)

live part

supplementary insulation

▼ **Figure 11.10**
Class II equipment with unearthed metal separated from live parts by basic and supplementary insulation including air gaps

11.2.7 Metal-encased Class II equipment

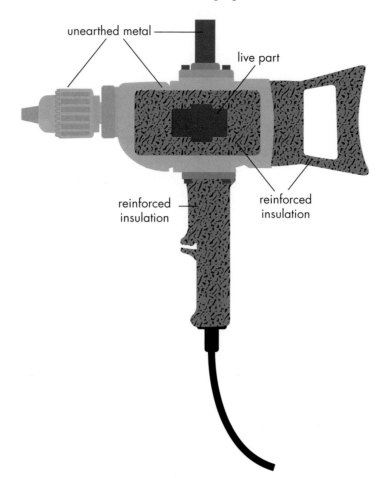

unearthed metal

live part

reinforced insulation

reinforced insulation

▼ **Figure 11.11**
Class II equipment
with a metal case that
is unearthed (Metal-
encased Class II)

Class II equipment should be identified with the Class II construction mark as shown in Figure 11.12.

▼ **Figure 11.12**
The Class II construction
mark

Notes
If an appliance with double insulation or reinforced insulation throughout has provision for earthing, it is considered to be a Class I or Class 0I appliance. See Section 11.4.2 for an explanation of Class 0I.

Class II appliances may incorporate means for maintaining the continuity of protective circuits, provided that such means are within the appliance and are insulated from conductive accessible parts by supplementary insulation.

11.3 Class III

Class III equipment relies for protection against electric shock on supply from a SELV source.

SELV is Separated Extra-Low Voltage. However, SELV is described as Safety Extra-Low Voltage in appliance standards, e.g. BS EN 60335, and Separated Extra-Low Voltage in installation standards, e.g. BS 7671.

The Class III construction mark is as shown in Figure 11.13.

SELV sources will not exceed 50 V a.c. and in many installations will be required to be below 24 V or 12 V. SELV systems require specialist design and it is a requirement that there is no earth facility in the distribution of a SELV circuit or on the appliance or equipment.

Class III equipment is required to be supplied from a safety isolating transformer to BS EN 60742 or BS EN 61558-2-6.

The safety isolating transformer will have the identification mark upon it as shown in Figure 11.14.

▼ **Figure 11.14**
The identification mark
for a safety isolating
transformer

The output winding of the safety isolating transformer is electrically separated from the input winding by insulation at least equivalent to double insulation or reinforced insulation.

Class III equipment may be encountered as SELV lighting for shops and offices.

11.4 Class 0 and 0I

Class 0 and Class 0I equipment is allowed only in very specific locations. Such equipment should not be used in the normal commercial, industrial or domestic environment. For completeness, an explanation of these two types of appliance is included below.

11.4.1 Class 0 equipment

As defined in Chapter 2, a Class 0 item of equipment has protection against electric shock that relies upon basic insulation only. There is no supplementary or reinforced insulation. There is no provision for the connection of accessible conductive parts, if any, to the protective conductor in the fixed wiring of the installation. In the event of failure of the basic insulation reliance is placed upon the environment in which the equipment is installed. An example of Class 0 equipment is certain older style mains powered series-connected Christmas tree lights where the lights are interconnected by a bell flex type cable (insulated but not sheathed).

11.4.2 Class 0I equipment

A Class 0I item of equipment has at least basic insulation throughout. The equipment is provided with an earthing terminal. The equipment has a power supply cord without a protective earthing conductor. The equipment is fitted with a plug without an earthing contact, which cannot be introduced into a socket-outlet with an earthing contact.

Class 0I equipment is specialist equipment and not for common use.

12 Types of inspection and testing

Three types of inspection and testing are recommended in this Code of Practice and are described in Table 7.1 (Initial Frequency of Inspection and Testing of Equipment) as follows:

- ▶ user checks (no record is made if the equipment is found to be satisfactory)
- ▶ formal visual inspections (recorded)
- ▶ combined inspection and testing (recorded)

These three types of inspection and testing are covered in Chapters 13, 14 and 15.

The user check   13

The user check is a vital safety precaution. Many faults can be determined by a visual inspection. The user is the person most familiar with the equipment and may be in the best position to know if it is in a safe condition and working properly. No record need be made of the user check unless some aspect of the equipment is unsatisfactory. Advice on the frequency of user checks is given in Table 7.1.

The user check is limited to an external visual inspection without any dismantling of the equipment, such as removal of covers or plug tops. Note that internal inspection of the item of equipment, involving dismantling as required, is undertaken as a part of the formal visual inspection outlined in Chapter 14 of this Code.

The user check should proceed as follows. The user should:

1 Consider whether he/she is aware of any fault in the equipment and whether it works properly.
2 Disconnect the equipment, if appropriate, by switching off and unplugging the item of equipment (Figure 13.1).
3 Inspect the equipment, the cable and the plug. The inspection should include the checks listed in Table 13.1.
4 Take action if any faults or damage are apparent. Faulty equipment should be
   i switched off and unplugged from the supply,
   ii labelled to identify that it is not to be used,
   iii reported to the responsible person, and
   iv removed from service as soon as possible.

If equipment is found to be damaged or faulty, an assessment should be made by a responsible person as to the suitability of the equipment for the use or the location.

Frequent inspections and tests will not prevent damage occurring if the equipment is unsuitable for the particular application. In this case, replacement by suitable equipment is required.

▼ **Figure 13.1**
The user should disconnect the equipment, if appropriate, by switching off and unplugging the item of equipment

▼ **Table 13.1**   User checks

Plug	Not loose in socket-outlet and can be removed from socket-outlet without difficulty
	Free from cracks or damage
	Free from any sign of overheating
	Flexible cable secure in its anchorage
	If the plug is of the non-rewireable type or moulded-on type, the cable grip should be checked by firmly pulling and twisting the cable. No movement should be apparent.
	Pins not bent
	Pins preferably sleeved, particularly where young children may touch the plug
	No cardboard label on the bottom
	Plug does not rattle
Flex or cable	Good condition
	Free from cuts, fraying and damage
	Not in a location where it could be damaged
	Not too long, too short or in any other way unsatisfactory
	No joints or connections that may render it unsuitable for use, such as taped joints
	Only one flex connected into one plug (a 13 A plug is designed for one cable only – not two)
	Not too tightly bent at any place
	Not run under a carpet
	Not a trip hazard
	An extension lead should be inspected throughout its length. This will mean uncoiling coiled-type extension leads (Figure 13.2).
Socket-outlet or flex outlet	Free from cracks or other damage
	No sign of overheating
	Shutter mechanism of socket-outlet functioning
	Not loose (i.e. properly secured)
	Switch, if fitted, operates correctly
Adaptor or extension lead fitted with an RCD	Inspect device and verify it has a rated residual operating current not exceeding 30 mA
	Check device by plugging it in, switching it on and then pushing the test button. The RCD should operate and disconnect the supply from the socket-outlet(s).
Appliance or item of equipment	Free from cracks, chemical or corrosion damage to the case, or damage that could result in access to live parts
	Equipment is operated with protective covers in place and doors closed
	Able to be used safely
	Switches on and off correctly
	Works properly
	Sufficient space to permit cooling. Not positioned so close to walls and partitions that there is inadequate spacing for ventilation and cooling.
	No sign of overheating (Figure 13.3)
	Not likely to overheat. No books or files on top of a computer or towels over a convector heater. 100 W lamps should not be fitted in a 60 W luminaire.
	Cups and plants are not placed where their contents could spill into equipment
Environment	Equipment suitable for its environment
	No indiscriminate use of extension leads or multiway adapters
	Equipment normally not left on overnight
Suitability	Equipment suitable for the work it is required to carry out

Note: See Appendix VIII for further details and diagrams

▼ **Figure 13.2**
Extension leads must
be uncoiled and
the extension lead
inspected throughout its
length

▼ **Figure 13.3**
A risk of fire exists if a
100 W lamp is fitted in
a luminaire rated for a
maximum of 60 W

The formal visual inspection

14

Only a person competent to do so should carry out a formal visual inspection.

The results of a formal visual inspection should be recorded on a form such as Form VI.2 in Appendix VI.

Advice on the initial frequency of formal visual inspections is given in Table 7.1.

14.1 Manufacturer's instructions

The equipment should be installed and operated in accordance with the manufacturer's instructions.

The correct voltage, frequency and current requirements should be verified.

Requirements or recommendations applicable to heat dissipation should be met.

The fuse recommended by the manufacturer should be fitted. A larger fuse should never be fitted. Fitting a smaller fuse, such as a 3 A fuse, may give problems later. For example, many items of equipment have considerable inrush currents when first energized due to the starting surge of motors or the charging currents of capacitors.

14.2 Suitability of the equipment for the environment

The inspector or test operative should consider whether the equipment being inspected and/or tested is suitable for:

► the environment; or
► the nature of the work being undertaken.

Equipment that is unsuitable for the environment or the nature of the work being undertaken should be recorded on the documentation and brought to the attention of the responsible person.

Particular care needs to be taken when selecting the equipment and assessing the frequency of inspection and testing when the work environment is harsh or hazardous and if the equipment is exposed to:

► mechanical damage
► the weather
► natural hazards
► high or low temperatures
► pressure
► water
► dirty conditions

- corrosive conditions
- flammable or explosive substances

In hazardous and particularly difficult environments, specialist advice needs to be taken and reference should be made to British Standards and Health and Safety Executive Guidance, e.g. the guidance on Regulation 6 in the Memorandum of Guidance on the Electricity at Work Regulations 1989 (see Appendix II).

14.3   Switching of equipment

The inspector or test operative should establish that there is a suitable means for switching the equipment.

The inspector should determine that, where needed, there is a means for switching off

1   for normal functional use,
2   to carry out maintenance, and
3   in an emergency (if required).

14.3.1   Functional switching

The means of switching off, the functional switch, should be readily accessible to the user, i.e. in normal circumstances it should be possible to reach the functional switch without difficulty.

14.3.2   Isolation and switching off for mechanical maintenance

The equipment should be able to be isolated from the supply. This will be simple to achieve when the equipment is connected via a plug and socket-outlet. However, some equipment may be connected to the supply by other means such as an isolator or connection unit, where isolation from the supply can be achieved only by switching off or by removing the fuse. The means of isolation may be at the consumer unit or distribution board. Refer to Appendix IX for further information. Great care should be taken when carrying out a visual inspection of equipment that does not have a visible means of isolation, as it may still be live with the attendant risk of electric shock.

Before switching off business, telecommunication and other equipment, the permission of the responsible person will need to be obtained, otherwise disruption or serious loss to the business may result. Similarly, the permission of the responsible person will need to be obtained before disconnecting communication links.

It should be noted that business equipment may need to be powered down before being switched off. Specific procedures may need to be followed to prevent damage or loss of data.

Equipment supplied via an uninterruptible power supply (or other standby supply) should be isolated from its standby source before the inspection commences.

14.3.3   Emergency switching

The purpose of emergency switching is largely explained by its definition, although it should be added that the purpose is not only to remove danger, but also to prevent danger that is perceived to be imminent. It should be noted that the term 'danger', as used in the definition of emergency switching and generally throughout BS 7671, has the following specific meaning:

Danger. Risk of injury to persons (and livestock where expected to be present) from

1 fire, electric shock and burns arising from the use of electrical energy; and
2 mechanical movement of electrically controlled equipment, in so far as such danger is intended to be prevented by electrical emergency switching or by electrical switching for mechanical maintenance of non-electrical parts of such equipment.

Thus, the purpose of emergency switching, expressed more fully, is to prevent or remove the risk of injury arising from the particular types of hazard referred to in 1 and 2 above. Emergency stopping has the additional function of stopping the operation (normally the mechanical movement) of electrically actuated equipment.

14.4 User feedback

Before inspecting the equipment, where possible, the user should be asked whether he/she is aware of any faults and whether the equipment works properly, and then the inspector proceeds accordingly. The user may be familiar with the equipment and may be aware of intermittent faults.

14.5 The equipment

The person performing the formal visual inspection should perform the User Checks detailed in Table 13.1. The inspector should remove the top from the plug, unless it is a moulded plug, and inspect the connections within. The connections should be checked for tightness. The fuse should be checked. The fuse in a moulded plug should be checked.

14.6 Equipment failing the formal visual inspection

If equipment is found to be damaged or faulty, an assessment should be made by a responsible person as to the suitability of the equipment for the use/location. If the responsible person concludes that the equipment is unsafe, it should be immediately removed from use and labelled. Repair should be arranged.

Note that frequent inspections and tests will not prevent damage occurring if the equipment is unsuitable for the particular application. Replacement by suitable equipment is required.

14.7 Recording the formal visual inspection

The formal visual inspection should be recorded and a model form (VI.2) is shown in Appendix VI.

Combined inspection and testing | 15

Safety

The most important safety issues when testing are:

1 Test equipment should be, so far as is reasonably practicable, constructed, maintained and used in such a way as to prevent danger.

2 Live working is not permitted unless it is
   i   unreasonable to work dead,
   ii it is reasonable to work live, and
   iii suitable precautions are taken to prevent injury.

3 Work is required to be carried out in a safe manner. Factors to consider when developing safe working practices include
   i   control of risks while working,
   ii   control of test areas,
   iii use of suitable tools and clothing,
   iv   use of suitable insulated barriers,
   v   adequate information,
   vi   adequate accompaniment, and
   vii adequate space, access and lighting.

4 People at work are required to prevent danger and injury, have adequate training, skill and experience and have adequate supervision when appropriate.

GS 38 (HSE publication *Electrical test equipment for use by electricians*) gives advice to competent persons on the selection and use of test equipment. Refer also to Section 10.1 of this publication for details of test probes.

When electrical testing is required it should be performed by a person who is competent in the safe use of the test equipment and who knows how to interpret the results obtained. This person should be capable of inspecting the equipment and, where necessary, dismantling it to check the cable connections. Care should always be exercised when conducting tests. Inappropriate tests can damage equipment.

If equipment is permanently connected to the fixed installation, e.g. by a flex outlet or other accessory, the accessory will need to be detached from its box or enclosure so that the connections can be inspected. Such work should only be carried out after isolation by a competent person.

The accessible conductive parts of the appliance should not be touched during any powered test such as the protective conductor/touch current test.

15

15.1 Preliminary inspection

Guidance on the initial frequency of inspection and testing is given in Table 7.1.

If the Class of equipment is not known, it is to be tested as Class I.

Before inspection and testing are carried out, the test operative should obtain a copy of the previous test results, if available, so that any deterioration can be assessed and advice given accordingly.

Before equipment is tested it is necessary to carry out a preliminary visual inspection. This visual inspection is the most important activity to be carried out on a piece of equipment. Often, formal testing with instruments will not indicate a failure that is apparent on inspection. For example, a damaged case exposing live parts will often not be apparent from an insulation test.

The preliminary inspection procedure is as follows.

1 Determine whether the equipment can be disconnected from the supply and disconnect if, and only if, permission is received. If permission is not received to disconnect the supply do not proceed with any tests. Record that the equipment has not been inspected or tested or has had only a minimal inspection and no tests, and label and report accordingly.
2 Disconnect business equipment from communication links after, and only after, receiving permission. Do not test equipment connected to communication links. Care should be taken with optical fibre systems to avoid exposure to invisible IR radiation. Do not hold up optical fibre links to the eye. Ensure that disconnected fibre ends are protected with their resident dust-caps.
3 Thoroughly inspect the appliance for signs of damage, as for a formal visual inspection (Chapter 14).
4 Inspect the plug as for a formal visual inspection (Chapter 14).
5 Inspect the flexible cable for damage throughout its length both by visual inspection and touch. Where extension leads are necessary, the guidance given in Section 15.10 should be followed.
6 Assess if the equipment is suitable for the environment.
7 Inspect the socket-outlet or flex outlet, as for a formal visual inspection (Chapter 14).

Further information is given in Appendix VIII.

15.2 Test procedures

Equipment that is supplied by a plug and socket-outlet connection can be readily tested by a dedicated portable appliance test instrument, by unplugging the equipment to be tested and plugging it into the dedicated test equipment.

Equipment that is permanently connected to a flex outlet type of accessory can more easily be tested using an insulation/continuity test instrument with the test leads connected directly to the accessory terminals. The supply to the accessory is required to be isolated and proved dead at the point of work before the testing commences. Note that if the apparatus has not been physically disconnected, the isolation device will also be tested with the apparatus.

15.3 In-service tests

In-service testing requires the following:

- ▶ an earth continuity test on Class I equipment
- ▶ insulation resistance testing, if applicable, or protective conductor current/touch current test or substitute/alternative leakage test (see Section 10.2)
- ▶ functional checks

The sequence of testing should always be: 1) earth continuity test; 2) insulation resistance test; 3) protective conductor current and/or touch current test or substitute/ alternative leakage test; and 4) functional test.

An insulation test should be carried out before any powered tests as it may detect a potentially dangerous insulation failure. Only then should other tests such as a protective conductor or touch current be performed, if appropriate.

Insulation resistance testing may be substituted by a protective conductor current/touch current measurement where insulation resistance testing is not appropriate due to the possibility of the item of equipment being damaged by the test voltage. Alternatively insulation resistance testing may be inappropriate because of voltage limiting devices in the equipment under test or the fact that the equipment under test has an electronic switch that requires mains power to close the switch.

Test operatives should be aware that it may not be possible to perform some of the tests prescribed by this Code of Practice, for example, electrical equipment should not be subjected to test voltages and currents that would result in damage. Some electrical test devices apply tests that are inappropriate and may even damage equipment containing electronic circuits, possibly causing degradation to safety. In particular, although this Code of Practice includes insulation resistance tests, equipment should not normally be subjected to dielectric strength testing (also known as flash testing or hi-pot testing) because this may damage insulation and may also indirectly damage low voltage electronic circuits unless appropriate precautions are taken.

15.4 The earth continuity test

This test can only be applied to Class I equipment or cords, that is equipment which

1. relies on a connection with Earth for its safety (protective earthing), and/or
2. needs a connection with Earth for it to work (functional earthing).

If protective earthing is being provided, as is likely for many household appliances, tools and luminaires, the earth continuity test is most important, because the safety of the appliance depends upon a good connection with the means of earthing of the fixed electrical installation.

One of the following two tests should be carried out.

- ▶ **'Hard' test.** A continuity measurement should be made with a test current not less than 1.5 times the rating of the fuse up to a maximum of the order of 26 A for a period of between 5 and 20 s. The continuity test should be made between accessible earthed metal parts of the equipment and the earth pin of the plug (or the earthing terminal of the fixed wiring supply). The resistance measurement should be observed while flexing the cable and an inspection of the flexible cable terminations at the equipment and the plug or flex outlet should be made. Any variation in the measured value should be investigated. The terminations should be inspected for any evidence of deterioration, poor contact, corrosion etc.

▶ **'Soft' test.** A continuity measurement should be made as above but with a short-circuit test current within the range 20mA to 200mA nominal. The continuity test should be made between accessible earthed metal parts of the equipment and the earth pin of the plug (or the earthing terminal of the fixed wiring supply). The resistance measurement should be observed while flexing the cable and an inspection of the flexible cable terminations at the equipment and the plug or flex outlet should be made. Any variation in the measured value should be investigated. The terminations should be inspected for any evidence of deterioration, poor contact, corrosion etc.

A continuity test should be made to all exposed-conductive-parts. This may require multiple continuity tests on a single appliance.

Care must be taken that alternative earth paths are not provided by inadvertent contact or connection to other equipment that may provide an earth path e.g. via a signal cable. This would result in erroneous measurements.

The measured resistance should not exceed the values given in Table 15.1.

▼ **Table 15.1**
Earth continuity readings

For appliances with a supply cord	$(0.1 + R)$ Ω where R is the resistance of the protective conductor of the supply cord
For appliances without a supply cord	0.1 Ω
For 3-core appliance cord sets (see also Section 15.9)	$(0.1 + R)$ Ω where R is the resistance of the protective conductor of the supply cord
For extension leads, multiway adaptors and RCD adaptors (see also Section 15.10)	$(0.1 + R)$ Ω where R is the resistance of the protective conductor of the supply cord

Notes:
1 Some equipment may have accessible metal parts that are earthed only for functional or screening purposes, with protection against electric shock being provided by double or reinforced insulation. It is very important that these non-safety earthed metal parts are not subjected to the 'hard test', otherwise damage may result. Connections may be checked using a low current continuity test instrument as in the 'soft test'.
2 Care should be taken to ensure that the contact resistance between the tip of the test probe and the metal part under test does not influence the test result.
3 The test should only be carried out for the duration necessary for a stable measurement to be made, and to allow time for flexing of the cable.
4 If the resistance R of the protective conductor of the supply cord cannot easily be measured, Table VII (in Appendix VII) provides nominal cable resistances per metre length for various types of cable. The supply cord csa should first be identified and the length measured. The resistance of the protective conductor can then be calculated.
5 Some portable appliance test instruments with go/no-go indication may fail cord connected appliances with earth continuity resistance exceeding 0.1 Ω. If it is not possible to re-programme the appliance test instrument it will be necessary for a measurement of the actual resistance to be made with another instrument.
6 A Class I appliance may have unearthed metal that is in casual or fortuitous contact with earthed metal as illustrated in Figure 11.4. A continuity test made to this 'unearthed' metal may give misleading test results. The 'unearthed' metal is not required to be earthed.

15.5 The insulation resistance test

Insulation resistance is normally measured by applying a test voltage of 500 V d.c. and measuring resistance. This test, a so-called 'hard test', may not always be suitable because it may damage IT equipment or other equipment containing electronic components. Therefore, a more appropriate alternative may be one of the following: the protective conductor/touch current measurement described in Section 15.6 (a so-called 'soft' test); or the substitute/alternative leakage current test (see Section 10.2); or the insulation resistance test at a reduced voltage such as 250 V d.c. (see Section 10.2).

Appliances should not be touched while carrying out insulation resistance tests as exposed metalwork may reach the test voltage, which, although not dangerous, could be uncomfortable and risk causing injury by involuntary movement.

The live conductors (phase and neutral) should be connected together for the insulation test. This is best achieved either by using special test equipment or by using a special test socket with the phase and neutral connected together. Equipment is not to be returned to service with any phase-neutral connections still in place. It is therefore recommended that the functional test be carried out last.

Before the test, the suitability of fuses in the equipment to be tested should be checked and power switches put in the ON position. All covers should be in place. The test is carried out between live conductors, i.e. phase and neutral, connected together, and the body of the appliance.

The applied test voltage should be approximately 500 V d.c. The test instrument should be capable of maintaining this test voltage with a load resistance of 0.5 MΩ. Insulation resistance readings obtained should be not less than the values shown in Table 15.2.

When testing insulation resistance on a Class II appliance the test probe should be connected to any metal parts or suspect joints in the enclosure where conductive material may have accumulated. This may require multiple tests.

▼ **Table 15.2**
Insulation resistance readings

Appliance class	Insulation resistance
Class I heating and cooking equipment with a rating ≥ 3 kW	0.3 MΩ
All other Class I equipment	1.0 MΩ
Class II equipment	2.0 MΩ
Class III equipment	250 kΩ

Notes:
1 Certain heating and cooking appliances may be unable to meet the insulation resistance requirements, such as where metal sheathed mineral-insulated heating elements are used. It may be necessary in some cases to switch on the appliance for a period of time to drive off absorbed moisture before commencing testing. Additionally the touch current measurement of Section 15.6 may be carried out.
2 It is important to ensure that the connections between the test instrument and the equipment under test are properly made. A poor earth connection, particularly for appliances with relatively high protective conductor current, may cause a perceived 'electric shock' from the appliance frame, which although not dangerous, could be uncomfortable and risk causing injury by involuntary movement.

3   This test, which applies 500 V d.c. to the item of equipment under test, should not be applied to IT (information technology) equipment, unless such equipment complies with the requirements of BS EN 60950: *Specification for safety of information technology equipment including electrical business equipment*. Equipment not constructed to this standard may be damaged by this test.

4   The phase and neutral of the equipment should be securely connected together while making this test. This is best achieved by using pre-wired automatically configured test equipment or by plugging into a special test socket.

5   For three-phase equipment, all three phases and neutral (if applicable) are to be linked together while conducting this test.

6   Some equipment may have filter networks or transient suppression devices that could cause the insulation resistance to be less than specified. The manufacturer or supplier is to be consulted in these cases as to the acceptable value of measured insulation resistance. A protective conductor and/or touch current test should be performed in addition to the insulation test.

15.6   Protective conductor/touch current measurement

The protective conductor/touch current test is an additional or complementary test to the insulation test of Section 15.5, for use if the insulation test cannot be performed or gives suspect results. It is a so-called 'soft' test and can be used for IT equipment. An insulation test should be carried out before any powered test, such as the protective conductor/touch current test, because the insulation test may detect a potentially dangerous loss of insulation. If the insulation test gives a very low or zero reading, the reason should be investigated before any further tests are performed.

The protective conductor current/touch current is measured from the internal live parts to Earth for Class I equipment, or from the internal live parts to accessible surfaces of Class II equipment. For practical purposes, the test voltage is the supply voltage.

The equipment should be switched on and hence is operating for the time during which the measurement is made. Suitable precautions should be taken. For example, if the item of equipment under test is a kettle, it should be filled with water to avoid damage to the element. There are potential hazards that may arise during powered tests on a rotating/moving machine, e.g. electric drill.

Note that the accessible conductive parts of the appliance should not be touched during a protective conductor/touch current test.

The protective conductor current or touch current should be measured within 5 s after the application of the test voltage, the supply voltage, and should not exceed the values in Table 15.3.

Appliance class	Maximum protective conductor or touch current (Note 1)
Portable or hand-held Class I equipment	0.75 mA
Class I heating appliances	0.75 mA or 0.75 mA per kW, whichever is the greater, with a maximum of 5 mA
Other Class I equipment	3.5 mA
Class II equipment	0.25 mA
Class III equipment	0.5 mA

▼ **Table 15.3**
Measured protective conductor/touch current

Notes:

1 The values for maximum protective conductor or touch current given above are to be doubled if
   i the appliance has no control device other than a thermal cut-out, a thermostat without an 'off' position or an energy regulator without an 'off' position, or
   ii all control devices have an 'off' position with a contact opening of at least 3 mm and disconnection in each pole.
2 Equipment with a protective conductor current designed to exceed 3.5 mA should comply with the requirements of Section 15.12.
3 The nominal test voltage is:
   i 1.06 times rated voltage, or 1.06 times the upper limit of the rated voltage range, for appliances for d.c. only, for single-phase appliances and for three-phase appliances that are also suitable for single-phase supply, if the rated voltage or the upper limit of the rated voltage range does not exceed 250 V.
   ii 1.06 times rated line voltage divided by 1.73, or 1.06 times the upper limit of the rated voltage range divided by 1.73 for other three-phase appliances.

15.7 Functional checks

The functional check is the last of the tests to be performed and it is simply a check to ensure that the item of equipment is working properly. The item of equipment is energized and it should work normally.

The use of more sophisticated instruments may permit load testing, which is an effective way of determining whether there are certain faults in appliances. It is particularly useful for heating appliances and will identify whether one or more elements are open circuit.

15.8 Damaged or faulty equipment

If equipment is found to be damaged or faulty on inspection or test, an assessment should be made by a responsible person as to the suitability of the equipment for the use or the location.

Any items found to be faulty or defective by the test operative, such as the hedge cutter illustrated in Figure 15.1, should be brought to the attention of the responsible person.

▼ Figure 15.1
Damaged or faulty equipment should be immediately removed from use and labelled

15.9 Appliance cord sets

An appliance with a detachable power supply flex (appliance-coupler) should be tested with the cord set plugged into the appliance.

The cord set should be labelled and then tested separately from the appliance as follows:

► a 3-core cord set should be tested as a Class I appliance
► a 2-core cord set should be tested as a Class II appliance

The following inspections and tests should be made:

► Visual inspection
► Class I – earth continuity, polarity and insulation checks
► Class II – polarity and insulation checks

The reason that the cord set is inspected and tested separately from the appliance is that the cord set could be used during the course of the next period to supply a different appliance. For example, if the cord set was 2-core and, during the year, was inadvertently used to supply a Class I appliance, the appliance would be unearthed and present a risk of electric shock.

A 2-core cord set should not be fitted with a 3-pole appliance coupler.

15.10 Extension leads, multiway adaptors and RCD adaptors

15.10.1 Extension leads

Extension leads should be checked and tested as follows.

1   The length of an extension lead should be checked to ensure that it is not so great that the appliance performance may be affected by voltage drop. Additionally, the length should not exceed the figures shown in Table 15.4.

csa of the core	Maximum length (metres)
1.25 mm^2	12
1.5 mm^2	15
2.5 mm^2	25

2.5 mm^2 extension leads are too large for standard 13 A plugs, although they may be used with BS EN 60309 industrial plugs.

Any extension lead exceeding the above lengths should be fitted with an RCD with a rated residual operating current not exceeding 30 mA to ensure that the lead and apparatus are protected in the event of an earth fault (i.e. fault protection is provided that is protection against electric shock by indirect contact) and for when the lead is used to supply portable equipment outdoors. However, the equipment supplied may not function correctly due to voltage drop in the cable. There may also be a risk of fire due to overloading and under fault conditions automatic disconnection may not occur within the prescribed time.

2 Where an extension lead is fitted with a standard 3-pin socket-outlet it should be tested as a Class I appliance.

3 A polarity check should be performed to ensure that the phase and neutral conductors are not crossed.

15.10.2 RCD extension leads

Where an extension lead is fitted with an RCD, the RCD must have a rated residual operating current not exceeding 30 mA.

▶ **Basic check of RCD.** The RCD should be checked for correct operation by plugging it in, switching it on and then pushing the test button. The RCD should operate and disconnect the supply from the socket(s) on the extension lead.

▶ **Testing an RCD.** In addition to the basic check, it is recommended that an RCD be tested with an RCD test instrument. The instrument manufacturer's instructions should be followed and the device must operate within the times given in Table 15.5 for a residual operating test current of 30 mA.

	100% of rated residual operating current	Satisfactory result
RCDs to BS 4293 and RCD-protected socket-outlets to BS 7288	30 mA	Device should trip in not more than 200 ms
RCDs to BS EN 61008	30 mA	Device should trip in not more than 300 ms

15.10.3 Multiway adaptors and RCD adaptors

Sufficient socket-outlets should be provided so that multiway adapters are not necessary. Due to the increasing amount of computer and electronic equipment this is often not achievable in offices and some multiway adapters may be necessary. The person inspecting the installation should decide what is reasonable in terms of safety and report to the responsible person if excessive numbers of adapters are in use. Such adapters should not be 'daisy-chained'. New buildings should have sufficient socket-outlets so that multiway adapters are not necessary (Figure 15.2).

Adaptors fitted with an RCD should be checked and tested as in Section 15.10.2.

Certain adaptors, often cube adaptors, are unfused, meaning that it is possible to overload the adaptor. This would result in a fire risk.

15.11 Microwave ovens

When inspecting and testing, microwave ovens showing any signs of damage, distortion or corrosion should be rejected. Return damaged ovens to specialist repairers only.

Microwave leakage should be checked at appropriate intervals (Figure 15.3).

A functional check to ascertain that opening of the door results in a reliable interruption of the oven power should be carried out.

15.12 High protective conductor currents

There are particular requirements in BS 7671 for the earthing arrangements for equipment having high protective conductor currents.

It should be noted that equipment with a protective conductor current designed to exceed 3.5 mA should

1   be permanently wired to the fixed installation, or be supplied by an industrial plug and socket to BS EN 60309-2,
2   have internal protective conductors of not less than 1.0 mm² csa (see clause 5.2.5 of BS EN 60950), and
3   have a label bearing the following warning or similar wording fixed adjacent to the equipment primary power connection (see clause 5.1.7 of BS EN 60950).

> **WARNING: HIGH PROTECTIVE CONDUCTOR CURRENT**
>
> **Earth connection essential before connecting the supply**

Further precautions need to be taken for equipment with a protective conductor current exceeding 10 mA; see Section 607 of BS 7671. Similarly, for final circuits supplying a number of socket-outlets, where it is known or reasonably to be expected that the total protective conductor current in normal service will exceed 10 mA, additional precautions need to be taken.

Where an item of equipment or an appliance has a high protective conductor current, such as the variable speed drive in Figure 15.4 that includes a filter, an electric shock can be received from exposed-conductive-parts or the earth terminal if the appliance is not earthed. It is most important that appliances with high protective conductor currents are properly connected with Earth before any supply is connected.

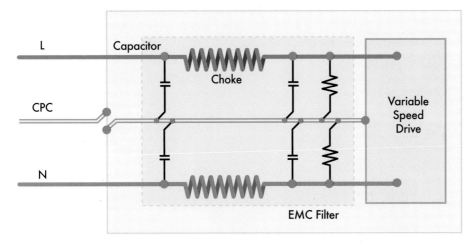

▼ **Figure 15.4**
A protective conductor current will result due to the capacitors and their discharge resistors in the EMC filter. Where appliances have high protective conductor currents, an electric shock can be received from exposed-conductive-parts or the earth terminal if the appliance is not earthed or the means of earthing is defective

15.13 Replacement of appliance flexes

The maximum lengths recommended for extension leads are not applicable to appliance flexes or cord sets.

For flexes to be protected by the fuse in a BS 1363 plug, there is no limit to their length, providing that their csas are as in Table 15.6.

Fuse rating	Minimum csa of cord or flex
3 A	0.5 mm²
13 A	1.25 mm²

Other considerations such as voltage drop may limit flex lengths. Smaller csas than those given above are acceptable if flex lengths are restricted. However, for replacement purposes the above simplified guidance is appropriate.

Refer also to Appendix VII.

15.14 Plug fuses

The fuse in the plug is not fitted to protect the appliance, although in practice it often does this. The fuse in the plug protects the flex against faults and can allow the use of a reduced csa flexible cable. This is advantageous for appliances such as electric blankets, soldering irons and Christmas tree lights, where the flexibility of a small flexible cable is desirable. Appliances are generally designed to European standards for use throughout Europe. In most countries the plug is unfused. If the appliance needs a fuse to comply with its manufacturing standard, it should be fitted in the appliance. For the convenience of users, appliance equipment manufacturers have standardized on two plug fuse ratings (3 A and 13 A) and adopted appropriate flex sizes. For appliances up to about 700 W, a 3 A fuse is generally used. For those over 700 W, a 13 A fuse may be used. However, note that, for some IT equipment, manufacturers fit 5 A fuses.

Although two standardized plug top fuse ratings have been adopted, the fuse recommended by the manufacturer should be fitted. Fitting a smaller fuse, such as a 3 A fuse, may give problems later. Many items of equipment have considerable inrush currents when first energized due to the starting surge of motors, the inrush currents of transformers or the charging currents of electronic power supplies.

Fuses should be to BS 1362 and have an ASTA mark as indicated in Figure 15.5.

Refer also to Appendix VIII.

▼ **Figure 15.5** A correctly wired plug top with a 13 A fuse to BS 1362 with the ASTA mark

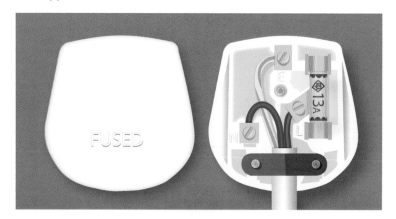

15.15 Equipment that cannot be located

Items of equipment on the register that the inspector or test operative is unable to locate should be brought to the attention of the responsible person.

Part 3
Appendices

British Standards

The following British Standards are relevant to equipment testing:

BS 1362: 1973 (1992)	Specification for general purpose fuse links for domestic and similar purposes (primarily for use in plugs).
BS 1363	13 A plugs, socket-outlets, connection units and adaptors.
BS 1363-1	1995 Specification for rewireable and non-rewireable 13 A fused plugs.
BS 1363-2	1995 Specification for 13 A switched and unswitched socket-outlets.
BS 1363-3	1995 Specification for adaptors.
BS 1363-4	1995 Specification for 13 A fused connection units: switched and unswitched.
BS 2754: 1976 (1999)	Memorandum. Construction of electrical equipment for protection against electric shock.
BS 5518: 1977 (1999)	Specification for electronic variable control switches (dimmer switches) for tungsten filament lighting.
BS 6004: 2000	Electric cables. PVC insulated, non-armoured cables for voltages up to and including 450/750 V, for electric power, lighting and internal wiring.
BS 6317	Specification for simple telephones for connection to public switched telephone networks run by certain public telecommunications operators.
BS 6360	Specification for conductors in insulated cables and cords.
BS 6450	Private branch exchanges for connection to public switched telephone networks run by certain public telecommunication operators.
BS 6500: 2000	Flexible cords rated up to 300/500 V, for use with appliances and equipment intended for domestic, office and similar environments.
BS 6972: 1988	Specification for general requirements for luminaire supporting couplers for domestic, light industrial and commercial use.
BS 7671: 2008	Requirements for Electrical Installations (The IEE Wiring Regulations).
BS 7895: 1997	Specification for bayonet lampholders with enhanced safety.
BS 7919: 2001	Electric cables. Flexible cables rated up to 450/750 V, for use with appliances and equipment intended for industrial and similar environments.
BS EN 50106	Specification for safety of household and similar electrical appliances. Particular rules for routine tests referring to appliances under the scope of BS EN 60335-1 and BS EN 60967.
BS EN 50144	Safety of hand-held electric motor operated tools.
BS EN 60051-1	Direct-acting electrical measuring instruments and their accessories.
BS EN 60065	Safety requirements for mains operated electronic and related apparatus for household and similar general use.
BS EN 60309	Plugs, socket-outlets and couplers for industrial purposes.
BS EN 60335	Safety of household and similar electrical appliances.
BS EN 60529	Degrees of protection provided by enclosures (IP code).

BS EN 60598	Safety of luminaires.
BS EN 60742	Isolating transformers and safety isolating transformers. Being replaced by BS EN 61558.
BS EN 60950	Specification for safety of information technology equipment including electrical business equipment. Identical to BS 7002.
BS EN 61008	Residual current operated circuit-breakers without integral overcurrent protection for household and similar uses (RCCBs).
BS EN 61009	Residual current operated circuit-breakers with integral overcurrent protection for household and similar uses (RCBOs).
BS EN 61010	Safety requirements for electrical equipment for measurement, control and laboratory use.
BS EN 61557	Electrical safety in low voltage distribution systems. Equipment for testing, measuring and monitoring of protective measures.
BS EN 61558	Safety of power transformers, power supply units and similar.
BS EN 61242	Electrical accessories. Cable reels for household and similar purposes.

British Standards are available from:

BSI Customer Services
389 Chiswick High Road
London W4 4AL

To order a hardcopy document or to follow up on an order placed, contact the customer services department:
Tel: +44 (0)20 8996 9001
Fax: +44 (0)20 8996 7001
Email: orders@bsi-global.com

To contact BSI Standards for other issues such as

▶ for help navigating the British Standards Online Website,
▶ to register on British Standards Online,
▶ in the event of difficulty placing an order on British Standards Online,
▶ if your membership number is not being recognized on British Standards Online,
▶ if your company is a subscriber to the British Standards Online Website and you cannot download Standards, or
▶ if you have forgotten your username and password,

contact the online support services department:
Tel: +44 (0)20 8996 7555
Fax: +44 (0)20 8996 7001
Email: bsonlinesupport@bsigroup.com

Legal references and notes

1	Health and Safety at Work etc. Act 1974	Statutory Instrument	Section II.1
2	The Electrical Equipment (Safety) Regulations 1994	SI 1994/3260 as amended	
3	The Plugs and Sockets etc. (Safety) Regulations 1994	SI 1994/1768	
4	The Electricity at Work Regulations 1989	SI 1989/635 as amended	Section II.2
5	The Management of Health and Safety at Work Regulations 1999	SI 1999/3242	Section II.3
6	Provision and Use of Work Equipment Regulations 1998	SI 1998/2306 as amended	Section II.4
7	The Lifting Operations and Lifting Equipment Regulations 1998	SI 1998/2307	
8	Health and Safety Commission and Health and Safety Executive Guidance and Approved Codes of Practice on meeting the requirements of the above:		

8 (continued):

  i L1 A Guide to the Health and Safety at Work etc. Act 1974.

  ii HSR25 Memorandum of Guidance on the Electricity at Work Regulations 1989.

  iii L21 Approved Code of Practice & Guidance on the Management of Health and Safety at Work Regulations 1999.

  iv L22 Approved Code of Practice & Guidance on the Provision and Use of Work Equipment Regulations 1998.

Items 1 to 8 are available from TSO Online Bookshop at www.tsoshop.co.uk. Contact TSO if you find you are having difficulty placing an order or if there is a problem with an order you have already placed at Customer Services by email at esupport@tso.co.uk or telephone +44 (0)870 242 2345. Hearing and speech impaired customers can also communicate via text telephone on +44 (0)870 240 3701.

TSO operates six bookshops: Belfast, Birmingham, Cardiff, Edinburgh, London and Manchester. These act as outlets for both TSO and client publications and as distribution hubs.

II.1 Health and Safety at Work etc. Act 1974

Section 2 of the Health and Safety at Work etc. Act 1974 puts on employers a general duty of care to their employees. Specifically:

It shall be the duty of every employer to ensure, so far as is reasonably practicable, the health, safety and welfare at work of all his employees.

(2) Without prejudice to the generality of an employer's duty under the preceding subsection, the matters to which that duty extends include in particular

(a) the provision and maintenance of plant and systems at work that are, so far as is reasonably practicable, safe and without risks to health; ....

Employers also have general duties to persons other than their employees as described in Section 3(1).

Section 7 of the Act imposes general duties on employees at work as follows:

It shall be the duty of every employee while at work

1 to take reasonable care for the health and safety of himself and other persons who may be affected by his acts or omissions at work; and
2 as regards any duty or requirement imposed on his employer or any other person by or under any of the relevant statutory provisions, to cooperate with him so far as is necessary to enable that duty or requirement to be performed or complied with.

The Health and Safety at Work etc. Act 1974 is all-embracing, requiring all those concerned with an undertaking to do all that is reasonable to ensure the health and safety not only of persons directly employed, but other persons who may be associated with the work undertaken by the business. The requirements of the Act are general and widely applicable.

II.2 The Electricity at Work Regulations 1989 Statutory Instrument No. 635

The purpose of the Electricity at Work Regulations 1989 is to prevent death or injury to anyone from any electrical cause as a result of, or in connection with, work activities.

The regulations impose duties upon employers, self-employed persons and employees while at work etc. Regulation 4 is so significant that it is worth quoting in full:

Systems, work activities and protective equipment.

1 All systems shall at all times be of such construction as to prevent, so far as is reasonably practicable, danger.
2 As may be necessary to prevent danger, all systems shall be maintained so as to prevent, so far as is reasonably practicable, such danger.

> **3** Every work activity, including operation, use and maintenance of a system and work near a system, shall be carried out in such a manner as not to give rise, so far as is reasonably practicable, to danger.
>
> **4** Any equipment provided under these regulations for the purpose of protecting persons at work on or near electrical equipment shall be suitable for the use for which it is provided, be maintained in a condition suitable for that use, and be properly used.

The Electricity at Work Regulations apply to all electrical equipment from battery hand lamps to 400 kV transmission lines. The source of energy, the distribution systems and the current consuming equipment are all covered.

The Regulations include specific requirements for the strength, capability, suitability, insulation, earthing, protection against excess current and isolation of electrical systems. They also extend to work and equipment associated with such systems, such as near overhead power lines. There are requirements for precautions to be taken before working on equipment made dead, and for work on or near live or charged conductors.

A Memorandum of Guidance on the Electricity at Work Regulations is available and includes the text of the Regulations as well as advice to aid compliance.

II.3 Management of Health and Safety at Work Regulations 1999 Statutory Instrument No. 3242

Regulation 3 of the Management of Health and Safety at Work Regulations 1999 requires all employers and self-employed persons to assess the risks to workers and others who may be affected by their undertaking. Employers with five or more employees should also record the significant findings of that assessment. Guidance on the Management of Health and Safety at Work Regulations 1999 can be found in the Approved Code of Practice & Guidance published by the Health and Safety Commission. The Code has special legal status; if you follow its advice you will be doing sufficient to comply with the law in respect of matters on which the Code gives advice. Failure to do so can lead to successful prosecution for breach of Health and Safety law. The Code is essential reading, not just for employers and managers but also for everyone with responsibility for other people in the work place (see Regulation 14(2)).

This legislation requires an organisation to:

1 Assess the risks to all persons associated with their electrical equipment, identifying the significant risks, e.g. portable equipment used out of doors, and make a record of the assessment.

2 As appropriate, appoint a competent person to take responsibility for electrical maintenance including inspection and testing, ensuring that the person given this responsibility is competent in that he or she has sufficient training and experience, knowledge and other qualities to enable him or her properly to support the organisation.

II.4 Provision and Use of Work Equipment Regulations 1998 Statutory Instrument No. 2306 as amended

II.4.1 General

The Provision and Use of Work Equipment Regulations 1998 (PUWER) cover most risks that can result from using work equipment. With respect to risks from electricity, compliance with the Electricity at Work Regulations 1989 (EAW Regulations) is likely to achieve compliance with PUWER Regulations 5-9, 19 and 22.

PUWER applies only to work equipment used by workers at work. This includes all work equipment (fixed, portable or transportable) connected to a source of electrical energy. PUWER does not apply to the fixed installations in buildings. The electrical safety of these installations is dealt with only by the EAW Regulations.

More detailed commentary on the individual requirements is given below.

II.4.2 Maintenance

Regulation 5(1) of PUWER requires employers to ensure that work equipment is maintained in an efficient state, in efficient working order and in good repair. This is consistent with the requirement in Regulation 4(2) of the EAW Regulations that systems shall be maintained to prevent danger.

Regulation 5(2) of PUWER requires employers to ensure that, where any machinery has a maintenance log, the log is kept up to date. Neither PUWER nor the EAW Regulations specifically require a maintenance log to be kept for machinery, but it is good practice to keep a log or record of maintenance. Records of visual checks of plugs, cables etc. carried out by users of electrical machinery would not normally be considered to constitute a maintenance log. However, if formal records of maintenance of electrical machinery are kept in a maintenance log, then that log shall be kept up to date.

II.4.3 Inspection of work equipment that poses electrical risks

Regulation 6 of PUWER introduces a specific requirement for the inspection of work equipment. The Approved Code of Practice & Guidance L22, *Safe Use of Work Equipment*, notes at paragraph 136 that:

> Where the risk assessment under Regulation 3 of the Management of Health and Safety at Work Regulations 1999 has identified a significant risk to the operator or other workers from the installation or use of work equipment, a suitable inspection should be carried out.

Note that, although the 1992 Regulations have since been replaced by the 1999 Regulations, the requirements for risk assessment have not changed.

Inspection is required for work equipment:

▶ After installation and before being put into service for the first time, or after assembly at a new site or in a new location.

▶ At suitable intervals where it is exposed to conditions causing deterioration, which is liable to result in dangerous situations, and each time that exceptional circumstances, which are liable to jeopardize the safety of the work equipment, have occurred.

The risk assessment carried out under the Management of Health and Safety at Work Regulations 1999 will determine whether there are significant electrical risks that might justify an inspection under PUWER. If there are significant electrical risks then a competent person, probably the person who is trained to inspect visually and maintain the electrical equipment under the EAW Regulations, is required to inspect the equipment and record the results of the inspection under PUWER.

The above requirement of Regulation 6 in PUWER is consistent with the requirement in Regulation 4(2) of the EAW Regulations that all electrical systems shall be maintained to prevent danger, so far as is reasonably practicable. Guidance on the EAW Regulations has always stressed the importance of inspection, testing and record keeping if justified by the risk. This is consistent with the more specific requirements of PUWER.

II.4.4 Specific risks

Regulation 7 of PUWER imposes requirements for the use, repair, modification, maintenance or servicing of work equipment in high risk situations, such as where a Permit to Work system may be appropriate. This is consistent with the requirements in Regulations 4(3), 13 and 14 of the EAW Regulations and guidance given previously on those Regulations.

II.4.5 Information and instructions

Regulation 8 of PUWER requires employers to provide to people who use work equipment adequate health and safety information and, where appropriate, written instructions about the equipment that they use. This duty extends to providing information and instructions about any electrical risks that are present. Details of what is needed are set out in the Regulation and the supporting guidance in the Approved Code of Practice L22. This is consistent with the requirements of Regulations 4(3) and 16 of the EAW Regulations.

II.4.6 Training

Regulation 9 of PUWER requires employers to ensure that all persons who use work equipment have received adequate training for purposes of health and safety. This is consistent with the requirements of Regulations 4(3) and 16 of the EAW Regulations.

II.4.7 Isolation from sources of energy

Regulation 19 of PUWER requires employers to ensure that, where appropriate, work equipment is provided with suitable means to isolate it from all sources of energy and to take appropriate measures to ensure that reconnection of any energy sources does not expose any person using work equipment to any risk to his/her health and safety. This is consistent with the requirements of Regulations 12 and 13 of the EAW Regulations.

II.4.8 Maintenance operations

Regulation 22 of PUWER requires employers to take appropriate measures to ensure that work equipment is so constructed or adapted that, so far as is reasonably practicable, maintenance operations that involve a risk to health and safety can be carried out while the equipment is shut down. Where this is not reasonably practicable, employers should ensure that maintenance operations can be carried out without exposing the person carrying them out to a risk to his/her health and safety, or appropriate measures should be taken for the protection of any person carrying out maintenance operations that involve a risk to his/her health and safety. This is consistent with the requirements of Regulations 4(3), 13 and 14 of the EAW Regulations.

The Electricity at Work Regulations

The principal Regulations in the Electricity at Work Regulations applicable to in-service testing are:

Regulation 4	Systems, work activities and protective equipment
Regulation 5	Strength and capability of electrical equipment
Regulation 6	Adverse or hazardous environments
Regulation 7	Insulation, protection and placing of conductors
Regulation 8	Earthing or other suitable precautions
Regulation 10	Connections
Regulation 12	Means of cutting off the supply and for isolation
Regulation 13	Precautions for work on equipment made dead
Regulation 14	Work on or near live conductors
Regulation 15	Working space, access and lighting
Regulation 16	Persons to be competent to prevent danger and injury

III.1 Regulation 4: Systems, work activities and protective equipment

III.1.1 Regulation 4(2)

As may be necessary to prevent danger, all systems shall be maintained so as to prevent, so far as is reasonably practicable, such danger.

A 'system' is an electrical system in which all the electrical equipment is, or may be, electrically connected to a common source of electrical energy and includes such source and such equipment. [Regulation 2(1)]

'Electrical equipment' includes anything used, intended to be used or installed for use, to generate, provide, transmit, transform, rectify, convert, conduct, distribute, control, store, measure or use electrical energy. [Regulation 2(1)]

III.1.2 Regulation 4(3)

> Every work activity, including operation, use and maintenance of a system and work near a system, shall be carried out in such a manner as not to give rise, so far as is reasonably practicable, to danger.

Work activities of any sort, whether directly or indirectly associated with an electrical system, should be carried out in a manner that, so far as is reasonably practicable, does not give rise to danger. The maintenance of electrical systems, which includes the inspection and testing, should be carried out only by persons competent to do so and the work itself should not put anyone at risk, including the person actually doing the work.

III.2 Regulation 5: Strength and capability of electrical equipment

> No electrical equipment shall be put into use where its strength and capability may be exceeded in such a way as may give rise to danger.

The regulation requires that before equipment is energized the characteristics of the system to which the equipment is connected are taken into account including those pertaining under normal conditions, possible transient conditions and prospective fault conditions, so that the equipment is not subjected to stress that it is not capable of handling without giving rise to danger. The effects to be considered include voltage stress and the heating and electromagnetic effects of current.

III.3 Regulation 6: Adverse or hazardous environments

> Electrical equipment that may reasonably foreseeably be exposed to:
>
> **(a)** mechanical damage;
> **(b)** the effects of weather, natural hazards, temperature or pressure;
> **(c)** the effects of wet, dirty, dusty or corrosive conditions; or
> **(d)** any flammable or explosive substance, including dust, vapours or gases
>
> shall be of such construction or as necessary protected as to prevent, so far as is reasonably practicable, danger arising from such exposure.

The regulation draws attention to the kinds of adverse conditions where danger could arise if equipment is not constructed and protected in order to withstand such exposure. The regulation requires that electrical equipment should be suitable for the environment and conditions of use to which it may reasonably foreseeably be exposed in order that danger that may arise from such exposure will be prevented, so far as is reasonably practicable. Particular attention should be paid to the IP rating (Index of Protection) of equipment. Guidance is also given on the use of reduced voltage systems on construction sites and elsewhere where particularly arduous or conducting locations may exist.

III.4 Regulation 7: Insulation, protection and placing of conductors

> All conductors in a system that may give rise to danger shall
>
> **(a)** be suitably covered with insulating material and as necessary protected so as to prevent, so far as is reasonably practicable, danger; or
>
> **(b)** have such precautions taken in respect of them (including, where appropriate, their being suitably placed) as will prevent, so far as is reasonably practicable, danger.

The regulation requires that danger be prevented, so far as is reasonably practicable, by the means detailed in either part (a) or (b). The danger to be protected against generally arises from differences in electrical potential (voltage) between circuit conductors or between such conductors and other conductors in a system – usually conductors at earth potential. The conventional approach is either to insulate the conductors or to so place them so that persons are unable to receive an electric shock or burn from these conductors.

III.5 Regulation 8: Earthing or other suitable precautions

> Precautions shall be taken, either by earthing or by other suitable means, to prevent danger arising when any conductor (other than a circuit conductor) which may reasonably foreseeably become charged as a result of either the use of a system, or a fault in a system, becomes so charged; and, for the purposes of ensuring compliance with this regulation, a conductor shall be regarded as earthed when it is connected to the general mass of Earth by conductors of sufficient strength and current-carrying capability to discharge electrical energy to Earth.

The regulation applies to any conductor, other than a circuit conductor, that is liable to become charged as a result of either the use of a system or a fault in a system. The regulation requires that precautions be taken to prevent danger resulting from that conductor becoming charged. Because the regulation applies to any conductor (other than circuit conductors), this may include the conductive parts of equipment, such as outer metallic casings, that can be touched, which although not live, may become live under fault conditions.

III.6 Regulation 10: Connections

> Where necessary to prevent danger, every joint and connection in a system shall be mechanically and electrically suitable for use.

The regulation requires that all connections in circuit and protective conductors, including connections to terminals, plugs and sockets, and any other means of joining or connecting conductors, should be suitable for the purposes for which they are used. This requirement applies equally to temporary and permanent connections. The insulation and conductance of the connections should be suitable, having regard to the conditions of use including likely fault conditions.

III.7 Regulation 12: Means of cutting off the supply and for isolation

> **(1)** Subject to paragraph (3), where necessary to prevent danger, suitable means (including, where appropriate, methods of identifying circuits) shall be available for
> **(a)** cutting off the supply of electrical energy to any electrical equipment; and
> **(b)** the isolation of any electrical equipment.
> **(2)** In paragraph (1), 'isolation' means the disconnection and separation of the electrical equipment from every source of electrical energy in such a way that this disconnection and separation is secure.
> **(3)** Paragraph (1) shall not apply to electrical equipment which is itself a source of electrical energy but, in such a case as is necessary, precautions shall be taken to prevent, so far as is reasonably practicable, danger.

The objective of this part of the regulation is to ensure that, where necessary to prevent danger, suitable means are available by which the electricity supply to any piece of equipment can be switched off. Switching can be, for example, by direct manual operation or by indirect operation via 'stop' buttons in control circuits of contactors or circuit-breakers. There may be a need to switch off electrical equipment for reasons other than preventing electrical danger but these considerations are outside the scope of the Regulations.

III.8 Regulation 13: Precautions for work on equipment made dead

> Adequate precautions shall be taken to prevent electrical equipment, which has been made dead in order to prevent danger while work is carried out on or near that equipment, from becoming electrically charged during that work if danger may thereby arise.

Regulation 13 relates to situations in which electrical equipment has been made dead in order that work either on it or near it may be carried out without danger. The regulation may apply during any work, be it electrical or non-electrical. The regulation requires adequate precautions to be taken to prevent the electrical equipment from becoming electrically charged, from whatever source, if this charging would give rise to danger. 'Charged' is discussed under Regulation 2.

III.9 Regulation 14: Work on or near live conductors

> No person shall be engaged in any work activity on or so near any live conductor (other than one suitably covered with insulating material so as to prevent danger) that danger may arise unless
>
> **(a)** it is unreasonable in all the circumstances for it to be dead; and
> **(b)** it is reasonable in all the circumstances for him/her to be at work on or near it while it is live; and
> **(c)** suitable precautions (including where necessary the provision of suitable protective equipment) are taken to prevent injury.

Live work includes live testing, for example the use of a potential indicator on mains power and control logic circuits. Regulation 14 will often apply to electrical testing. Testing to establish whether electrical conductors are live or dead should always be done on the assumption that they may be live and therefore it should be assumed that this regulation is applicable until such time as the conductors have been proved dead. When testing for confirmation of a 'dead' circuit, the test instrument or voltage indicator used for this purpose should itself be proved, preferably immediately before and immediately after testing the conductors. Although live testing may be justifiable it does not follow that there will necessarily be justification for subsequent repair work to be carried out live.

III.10 Regulation 15: Working space, access and lighting

> For the purposes of enabling injury to be prevented, adequate working space, adequate means of access, and adequate lighting shall be provided at all electrical equipment on which, or near which, work is being done in circumstances which may give rise to danger.

The purpose of this regulation is to ensure that sufficient space, access and adequate illumination are provided while persons are working on, at, or near electrical equipment in order that they may work safely.

III.11 Regulation 16: Persons to be competent to prevent danger and injury

> No person shall be engaged in any work activity where technical knowledge or experience is necessary to prevent danger or, where appropriate, injury unless he/she possesses such knowledge or experience, or is under such degree of supervision as may be appropriate having regard to the nature of the work.

The object of the regulation is to ensure that persons are not placed at risk due to a lack of skills on the part of themselves or others in dealing with electrical equipment.

'... prevent danger or, where appropriate, injury ...'

This regulation uses both of the terms 'injury' and 'danger'. The regulation therefore applies to the whole range of work associated with electrical equipment where danger may arise and whether or not danger (or the risk of injury) is actually present during the work. It will include situations where the elimination of the risk of injury, i.e. the prevention of danger, for the duration of the work is under the control of a person who must therefore possess sufficient technical knowledge or experience, or be so supervised, etc. to be capable of ensuring that danger is prevented. For example, where a person is to effect the isolation of some electrical equipment before this person undertakes some work on the equipment, they will require sufficient technical knowledge or experience to prevent danger during the isolation. There will be no danger from the equipment during the work provided that the isolation has been carried out properly; danger will have been prevented but the person doing the work must have sufficient technical knowledge or experience so as to prevent danger during that work, for example by knowing not to work on adjacent 'live' circuits.

But the regulation also covers those circumstances where danger is present, i.e. where there is a risk of injury, as for example where work is being done on live or charged equipment using special techniques and under the terms of Regulation 14. In these circumstances persons must possess sufficient technical knowledge or experience or be so supervised etc., to be capable of ensuring that injury is prevented.

Technical knowledge or experience

The scope of 'technical knowledge or experience' may include:

1   adequate knowledge of electricity,
2   adequate experience of electrical work,
3   adequate understanding of the system to be worked on and practical experience of that class of system,
4   understanding of the hazards which may arise during the work and the precautions which need to be taken,
5   ability to recognise at all times whether it is safe for work to continue.

Allocation of responsibilities

Employees should be trained and instructed to ensure that they understand the safety procedures relevant to their work and should work in accordance with any instructions or rules directed at ensuring safety which have been laid down by their employer.

Supervision

The regulation recognises that in many circumstances persons will require to be supervised to some degree where their technical knowledge or experience is not of itself sufficient to ensure that they can otherwise undertake the work safely. The responsibilities of those undertaking the supervision should be clearly stated to them by those duty holders who allocate the responsibilities for supervision and consideration should be given to stating these responsibilities in writing. Where the risks involved are low, verbal instructions are likely to be adequate but as the risk or complexity increase there comes a point where the need for written procedures becomes important in order that instructions may be understood and supervised more rigorously. In this context, supervision does not necessarily require continual attendance at the work site, but the degree of supervision and the manner in which it is exercised is for the duty holders to arrange to ensure that danger, or as the case may be, injury, is prevented.

Summary of legislation and guidance

The *Code of Practice* considers the obligations imposed on persons by Legislation (Table IV.1), BS 7671: *Requirements for Electrical Installations* (Table IV.2) and Guidance (Table IV.3). A list of Health and Safety Executive (HSE) publications on Electrical Safety is given in Table IV.4

▼ **Table IV.1**
Summary of legislation

	Statutory legislation	Guidance document	
1	Health and Safety at Work etc. Act 1974 (HSAW)		
2	Electricity at Work Regulations 1989	Memorandum of Guidance HSR 25	[Courtesy of the HSE]
3	Management of Health and Safety at Work Regulations 1999 (MHSWR)	Approved Code of Practice L21	[Courtesy of the HSE]
4	Workplace (Health, Safety and Welfare) Regulations 1992	Approved Code of Practice L24	[Courtesy of the HSE]
5	Provision and Use of Work Equipment Regulations 1998 (PUWER)	Approved Code of Practice L22	[Courtesy of the HSE]

▼ **Table IV.2**
BS 7671: *Requirements for Electrical Installations*

Requirements for the fixed installation including inspection and testing	BS 7671:2008 *Requirements for Electrical Installations (The IEE Wiring Regulations)*	

▼ **Table IV.3**
Guidance

1	Guidance on the testing of the fixed installation	Guidance Note 3: *Inspection and Testing*	
2	HSE Guidance on electrical safety on construction sites	HSG 141 (formerly GS 24) *Electrical safety on construction sites*	[Courtesy of the HSE]
3	HSE Guidance on the selection and use of test probes, leads, lamps, voltage indicating devices and measuring equipment	GS 38: *Electrical test equipment for use by electricians*	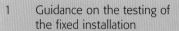[Courtesy of the HSE]
4	HSE Guidance on maintaining portable and transportable electrical equipment	HSG 107: *Maintaining portable and transportable electrical equipment*	[Courtesy of the HSE]
5	HSE Guidance on devising safe working practices for people working on or near electrical equipment	HSG 85: *Electricity at work, safe working practices*	[Courtesy of the HSE]

▼ **Table IV.4**   Health and Safety Executive (HSE) publications on Electrical Safety

Guidance document	Title	Electricity at Work Regulations particularly relevant
HS(R) 25	Memorandum of Guidance on the Electricity at Work Regulations 1989 1989 ISBN 07176 1602 9	
Video	Live wires. What to look for when inspecting portable equipment 1994 ISBN 0 7176 1916 8	4, 6, 7, 8, 9, 10, 11 and 12
CD-ROM	Your guide to the essentials of electrical safety 2000 ISBN 0 7176 1714 9	
EIS35 Free single copy	Safety in electrical testing; Servicing and repair of domestic appliances 2002	
EIS36 Free single copy	Safety in electrical testing; Servicing and repair of audio, TV and computer equipment 2002	
GS 6	Avoidance of danger from overhead electric power lines 1997 ISBN 0 7176 1348 8	4, 14, 15 and 16
GS 38	Electrical test equipment for use by electricians 1995 ISBN 0 7176 0845 X	4, 5, 6, 7, 10, 14 and 16
GS 50	Electrical safety at places of entertainment 1997 ISBN 07176 1387 9	
HSG 47	Avoiding danger from underground services 2000 ISBN 0 7176 1000 4	4, 14 and 16
HSG 85	Electricity at work: Safe working practices 2003 ISBN 0 7176 2164 2	
HSG107	Maintaining portable and transportable electrical equipment 1994 ISBN 0 7176 0715 1	4, 6, 7, 8, 9, 10, 11 and 12
HSG 118	Electrical safety in arc welding 1994 ISBN 0 7176 0704 6	4, 6, 7, 8, 10 12, 14 and 16
HSG 141 (formerly GS 24)	Electrical safety on construction sites 1995 ISBN 0 7176 1000 4	4–16 inclusive
INDG68 Free single copy	Do you use a steam/water pressure cleaner? You could be in for a shock. 1997	
INDG139 Free single copy	Electric storage batteries. Safe charging and use. 1993	
INDG231 Free single copy	Electrical safety and you 1996	

INDG236 Free single copy	Maintaining portable electrical equipment in offices and other low risk environments 1996 ISBN 0 7176 1272 4	4, 6, 7, 8, 9, 10, 11 and 12
INDG237 Free single copy	Maintaining portable electrical equipment in hotels and tourist accommodation 1996 ISBN 0 7176 1273 2	4, 6, 7, 8, 9, 10, 11 and 12
INDG247 Free single copy	Electrical safety for entertainers 1997 ISBN 07176 1406 9	
INDG354 Free single copy	Safety in electrical testing at work 1997 ISBN 07176 2296 7	
INDG372 Free single copy	Electrical switchgear and safety. A concise guide for users 2003 ISBN 0 7176 2187 1	
INDG389 Free single copy	Shock horror. Safe working near overhead power lines in agriculture 2003 ISBN 0 7176 2187 1	
PM 29	Electrical hazards from steam/water pressure cleaners 1995 ISBN 0 7176 0813 1	4, 6, 7, 8 and 10

The above publications are available from:

HSE Books, PO Box 1999, Sudbury, Suffolk, CO10 2WA
Tel: +44 (0)1787 881165
Fax: +44 (0)1787 313995
Email: hsebooks@prolog.uk.com
Web: www.hsebooks.com

HSE publications are also available through good booksellers.

Further information is available from:

HSE Infoline: +44 (0)845 345 0055
Email: hseinformationservices@natbrit.com
Web: www.hse.gov.uk

Production testing

This Appendix gives information on manufacturers' production testing.

Manufacturers of equipment will carry out tests on equipment they manufacture to check that the equipment has been manufactured as intended.

Where equipment is approved by an approval body, e.g. BEAB, BABT, BSI, the approval body will agree with the manufacturer the inspection, test and quality assurance procedures to be followed to ensure that the products are safe, and within accepted manufacturing tolerance of the samples type-tested.

BEAB refers to the ASTA BEAB Certification Services Organisation, which offers electrical approval, certification, testing and quality services to the electrical, electronic and allied trades.

BABT certifies products and services in the fields of radio and telecommunications, providing certification schemes and giving confidence that products and services comply with objective standards.

The tests described in this Appendix are typical of the electrical safety tests that may be required.

These tests are included for reference purposes only to assist in determining, in cases of doubt, whether the results of in-service tests are acceptable. In-service test results would not be expected to be 'better' than manufacturing test results. It is to be noted that in-service tests are not necessarily the same as manufacturing tests with respect to applied test voltages and currents.

Manufacturing production tests may be relatively arduous and generally should be applied only to equipment in as-new condition. They are reproduced here because they may well be useful for those testing as-new equipment or repaired equipment if appropriate.

V.1 BEAB for BS EN 60335 series, household and similar appliances

BEAB Document 40, Test Parameters for Appliances covered by the BS EN 60335 series of Standards.

References
BS EN 60335-1: 1994 *Safety of Household Electrical Appliances* and BS EN 50106: 1997 *Routine Tests for Household Appliances*.

V.1.1 Introduction

The tests defined in the BEAB document are intended to reveal a variation during manufacture that could impair the safety of household electrical appliances whose construction is covered by Harmonised Standard BS EN 60335. The tests do not impair the product's properties or reliability and are to be performed on every appliance unless otherwise agreed by BEAB or its authorized representative. Testing is normally carried out on the complete product after assembly with only labelling and packaging being performed after the final safety test. However, the manufacturer may perform the tests at a more appropriate stage in production provided that BEAB agrees that the later manufacturing operations will not affect the results.

If a flexible cord is provided, the appliance should be tested with the cord fitted. If an appliance fails any of the tests below it should be subjected to all the tests following repair and/or adjustment. The tests specified below are similar to those specified in Harmonised Standard BS EN 50106 but in some cases BEAB recommends different test values in order to ensure that appropriate safety levels are maintained. The specified tests are the minimum tests considered necessary to cover essential safety aspects. It is the responsibility of the manufacturer to decide if additional routine tests are necessary (see Part 1, Paragraph 8). Following completion of the functional tests the appliance should be subjected to the electrical safety tests with the appliance switched ON.

V.1.2 Functional tests and load deviation

All appliances should be subjected to a function test to verify correct operation. Any abnormal or out-of-limits results should be fully investigated in order to ensure safety will not be affected. The functioning of a component is checked by inspection or an appropriate test if a malfunction could result in a hazard. Verifying the direction of rotation of motors or the appropriate operation of switches and controls are examples of checks that may be necessary.

BEAB recommends:

> The power input of every appliance should be measured to ensure compliance with the appropriate Standard to which the appliance has been approved.

V.1.3 Earth continuity test

For Class I appliances, a current of at least 10A (BEAB recommends 25 A) derived from a source having a no-load voltage not exceeding 12 V, is passed between each of the accessible earthed metal parts and either of the following points:

▶ the earthing pin or earthing contact of the supply cord plug
▶ the earthing terminal of appliances intended to be connected to fixed wiring
▶ the earthing contact of the appliance inlet

The factory-applied limit for appliances with a supply cord is not to exceed $0.2 \, \Omega$ $((0.1 + R) \, \Omega$, where R = the resistance of the supply cord).

For all other appliances the resistance should not exceed $0.1 \, \Omega$.

The test is only carried out for the duration necessary for the measurement to be made; this is particularly important where the nominal csa of the cables is less than $0.75 \, \text{mm}^2$.

BEAB recommends:

> The test of the Standard (carried out with a current of 25 A) is designed to determine two things:
>
> 1 That the earthing path has a sufficiently low resistance;
> 2 That the fuse in the appliance plug or the supply circuit will blow before a weak point in the earthing path (e.g. a path made by a single-strand connection).

The test specified in BS EN 50106 will check point 1 but not 2. It is BEAB's view that the hazard presented by a lack of integrity in the earthing path should be detected in routine production tests, and that the test of the Safety Standard should be applied. There is no evidence that using a current of 25 A damages the appliance, but if a manufacturer has an appliance where this is a concern, then a lower value can be considered for that specific case. It is BEAB's experience that this is very seldom necessary.

V.1.4   Electric strength tests

The insulation of the appliance is subjected to a voltage of substantially sinusoidal waveform having a frequency of 50 Hz or 60 Hz for 1 s. No breakdown or flash-over should occur during the test. The value of the test voltage (rms) and the points of application are shown in Table V.1.

| Points of application | Test voltage V | | |
	Class I appliances	Class II appliances	Class III appliances
1 Between live parts and accessible metal parts separated from live parts by			
▶ Basic insulation only	1000 (1250)		
▶ Double or reinforced insulation	2500	2500	400 (500)
2 Between live parts and metal parts separated from live parts by basic insulation only		1000 (1250)	
3 Between inaccessible metal parts and the body		(2500)	

▼ **Table V.1**
Electric strength tests

Notes:
A d.c. test voltage may be used instead of a.c.; the values of the d.c. test voltages should then be 1.5 times those shown in the table.
BEAB recommends using the values shown in italics.

The insulation resistance test and dielectric strength test may be combined into a single test by using an electric strength test set that incorporates a current sensing device which normally trips when the current exceeds 5 mA. Tripping the sensing device should activate an audible or visual indication of breakdown of the insulation. The high voltage transformer should be capable of maintaining the specified voltage until the tripping current flows.

BEAB recommends:

> The leakage (touch) current limit should be set to the equivalent minimum insulation resistance requirement for the product as given in the Standard (e.g. with a test voltage of 3750 V rms and an insulation resistance of 7 megohms, the leakage (touch) current limit will be 0.54 mA rms).

V.1.5 Additional requirements for microwave ovens covered by the BS EN 60335-2-25 safety standard

Electric Strength Test – the Electric strength test equipment may have the current sensing device set up to 100 mA.

Check that the warnings concerning microwave energy specified in BS EN 60335-2-25 are marked on the relevant covers.

Check that the appliance is provided with the correct instructions for that particular appliance.

Check the operation of the door interlock system to ensure that microwave generation ceases when the door is opened.

The microwave oven is operated at rated voltage and with microwave power set to maximum. The energy flux density of microwave leakage is measured at any point approximately 5 cm from the external surface of the appliance. An appropriate load may be used. The measuring instrument is moved over the surface of the oven to locate the points of maximum leakage, particular attention being given to the door and its seals.

The microwave leakage should not exceed 50 W/m^2.

BEAB recommends:

> (a) that the appliance is operated at the upper end of the voltage range;
> (b) that the test load should be 1000 g +/– 5 g of potable water or equivalent;
> (c) that the microwave leakage monitor be regularly calibrated (recommended frequency = monthly) against a standard leakage source which itself is calibrated annually. Records of all checks are to be made and these should be available for inspection.

V.2 BEAB for BS EN 60065, audio, video and similar appliances

BEAB Document 40, Test Parameters for Electronic Equipment covered by BS EN 60065.

References

BS EN 60065: 1998 *Audio, Video and Similar Electronic Apparatus – Safety Requirements*.

V.2.1 Introduction

These tests are intended to reveal a variation during manufacture that could impair the safety of household electronic apparatus connected to the mains supply for indoor use, the construction of which is covered by Harmonised Standard BS EN 60065. The tests do not impair the properties or reliability of the apparatus.

V.2.2 Functional tests

Functional or performance tests are to be carried out as considered necessary by the manufacturer prior to the final electrical safety test being performed.

V.2.3 General considerations

The electrical safety tests listed are normally to be carried out, as applicable, 100 per cent at the final stage of manufacture prior to packaging. However, the manufacturer may perform the tests at an appropriate stage during production providing that the BEAB inspector agrees that the later manufacturing operations will not affect the results.

V.2.4 Test method

For all tests, with the exception of the earth continuity test on Class I equipment, any mains switch should be in the ON position.

Due account should be taken of the charging time for the total capacitance of the product under test particularly when using d.c. test voltages.

Test failures are to be indicated by visual or audible means.

V.2.5 Earth continuity test

Products of Class I construction that have accessible metal parts which may become live in the event of an insulation fault should be tested as follows.

A current of 10A (BEAB recommendation 25 A), derived from an a.c. source having a no-load voltage not exceeding 12V, is passed between those metal parts and the protective conductor of the mains supply cord for 3 s. The resistance should not exceed $(0.1 + R)$ Ω, where R is the resistance of the mains supply cord.

V.2.6 Insulation resistance

The insulation resistance test is to be measured on all products between both poles of the mains supply cord or terminals connected together and accessible metal parts. The test is applied using the voltage and limits shown in Table V.2.

▼ **Table V.2**
Test requirements for
electric strength tests

Type of test	Test value applied	Manufacturer's limit applied
1 Insulation resistance		**Minimum of**
Class I products	500 V d.c.	2 MΩ
Class II products	500 V d.c.	4 MΩ
2 Electric strength		
Basic insulation	1500 V rms	No flash-over
Supplementary insulation	2500 *(3000)* V rms	or breakdown
Reinforced insulation	2500 *(3000)* V rms	should occur

Notes:
1 Class II areas in Class I products are to be subjected to the electric strength test value for supplementary/reinforced insulation.
2 If the electric strength test equipment has an automatic indication of insulation breakdown that should be manually reset, the trip current setting should not normally exceed 6 mA rms.
3 BEAB recommends the use of the test values shown above in italics.

V.2.7   Electric strength test

This test is conducted between live parts, with both poles of the mains supply cord or terminals connected together, and accessible metal parts using a voltage of substantially sine wave form, having a frequency of 50/60 Hz and the value as shown in Table V.2, or a d.c. voltage equivalent to 1.414 times the rms voltage, applied for 6 s or for 3 s if the test equipment incorporates an automatic indication of insulation breakdown and has to be manually reset by the operator.

V.2.8   Combined test

The insulation resistance and electric strength tests may be combined using an electric strength test set incorporating a sensitive current trip set to show a fail result when it exceeds 0.75 mA at 1500 V rms (Class I) or 0.75 mA at 3000 V rms (Class II).

V.2.9   Specific requirements for television room aerials

Television room aerials should be subjected to safety tests covering insulation resistance and electric strength. Contact BEAB for further details of these tests.

V.3   BEAB for BS EN 60950, information technology and similar appliances

BEAB Document 40, *Test Parameters for Information Technology Equipment* covered by BS EN 60950

References
BS EN 60950: 1992 *Safety of Information Technology Equipment, Including Electrical Business Equipment*

V.3.1   Introduction

These tests are intended to reveal a variation during manufacture that could impair the safety of mains or battery powered IT equipment for home or business use, the construction of which equipment complies with Harmonised Standard BS EN 60950. The tests do not impair the properties or reliability of the apparatus.

V.3.2   Functional tests

The equipment should be subjected to functional or performance tests as considered necessary by the manufacturer prior to the final electrical safety test being performed.

V.3.3   General considerations

The electrical safety tests listed are to be carried out, as applicable, 100 per cent at the final stage of manufacture prior to packaging unless otherwise agreed. However, the manufacturer may perform the tests at an appropriate stage during production provided that the BEAB inspector agrees that the later manufacturing operations will not affect the results.

In addition to the requirements shown in Part 1, Certificates of Conformity for materials and components should confirm that the following applies.

1   High voltage components comply with BS EN 60950 Clause 4.4 or equivalent.
2   Capacitors that bridge insulation comply with BS EN 60950 Clauses 1.5.6 / 1.6.4 and IEC 60384 Part 14 (latest edition).

Plastic parts used in construction and printed circuit boards comply with the current Flame Retardancy Classification (FRC), such as granted by Underwriters Laboratories Inc.

V.3.4   Test method

For all tests, with the exception of the earth continuity test on Class I equipment, any mains switch should be in the ON position.

Due account should be taken of the charging time for the total capacitance of the product under test particularly when using d.c. test voltages. Note that the insulation resistance test of BS EN 60065 is replaced by an earth leakage test for BS EN 60950 products.

Test failures are to be indicated by visual or audible means.

V.3.5   Earth continuity test

Products of Class I construction that have accessible metal parts, which may become live in the event of an insulation fault, should be tested as follows.

A current of 10 A (BEAB recommendation 25 A), derived from an a.c. source having a no-load voltage not exceeding 12 V, is passed between those metal parts and the protective conductor of the mains supply cord for 3 s. The resistance should not exceed $(0.1 + R)\ \Omega$, where R is the resistance of the supply cord.

V.3.6   Earth leakage test

This test is carried out at the rated mains input voltage and the maximum leakage should not exceed the following.

► Class I products: 3.5 mA (0.75 mA for hand-held appliances)
► Class II products: 0.25 mA

V.3.7 Electric strength test

This test is conducted between live parts, with both poles of the mains supply cord or terminals connected together, and accessible metal parts using a voltage of substantially sine wave form, having a frequency of 50/60 Hz and the value shown in Table V.3, or a d.c. voltage equal to 1.414 times the rms voltage, applied for 6 s or for 3 s if the test equipment incorporates an automatic indication of insulation breakdown and has to be manually reset by the operator.

▼ **Table V.3**
Test requirements for electric strength tests

Type of test		Test value applied	Factory limit applied
1	**Earth leakage**		
	Class I: Hand-held appliances		0.75 mA
	Other Class I appliances	Rated input	3.5 mA
	Class II appliances		0.25 mA
2	**Electric strength**		
	Operational insulation	1500 V	No flash-over or breakdown should occur
	Basic insulation	1500 V	
	Supplementary insulation	1500 V	
	Reinforced insulation	3000 V	

Notes:
1 Insulation breakdown is considered to have occurred when the insulation does not restrict the uncontrolled flow of current.
2 When testing equipment incorporating solid state components that might be damaged by the secondary effect of the testing, the test may be conducted without the components being electrically connected providing that the wiring and terminal spacings are maintained.

V.3.8 Combined test

The earth leakage and electric strength tests may be combined using an electric strength test set incorporating a sensitive current trip set to show a fail result when it exceeds 20 mA for Class I and 3 mA at 3 kV for Class II products.

V.4 BABT

BABT Document 440, *Electrical and acoustic safety test* (see Table V.4).

▼ **Table V.4**   BABT electrical and acoustic safety tests

No.	Test connection points	Test title	Test condition	Test limits	Notes
1	Connection point 1: Main protective earth connection within equipment. Connection point 2: Other user accessible parts of equipment that have been connected to protective earth for safety reasons (and are hence protectively earthed).	Earth continuity	Max test voltage: 12 V a.c. or d.c. Min test current: 1.5 times current rating of the primary fuse Max test current 25 A	Measured resistance to be 0.1 Ω or less	1, 2, 3, 12–15
2	Connection point 1: Phase and neutral conductors shorted together. Connection point 2: Protective earth connection.	Electric strength for basic insulation	Test voltage: 1500 V a.c. or 2121 V d.c. Test time: 2 s min, 6 s max	No breakdown	4, 7, 8, 9, 10, 12–15
3	Connection point 1: NTP connectors shorted together. Connection point 2: Conductive parts separated from the NTP by basic or supplementary insulation, shorted together.	Electric strength for basic and supplementary insulation	Test voltage: 1500 V a.c. or 2121 V d.c. Test time: 2 s min, 6 s max	No breakdown	5, 7, 8, 10, 12–15
4	Connection point 1: Phase and neutral conductors shorted together. Connection point 2: Unearthed user accessible conductive parts or unearthed SELV outputs of a power supply shorted together.	Electric strength for reinforced insulation	Test voltage: 1500 V a.c. or 2121 V d.c. Test time: 2 s min, 6 s max	No breakdown	6, 7, 8, 9, 10, 12–15
5	Connection point 1: Phase and neutral conductors shorted together. Connection point 2: NTP connectors that are not protectively earthed, shorted together.	Electric strength for reinforced insulation	Test voltage: 3000 V a.c. or 4242 V d.c. Test time: 2 s min, 6 s max	No breakdown	6, 7, 8, 9, 10, 12–15
6	Connection point 1: NTP connectors shorted together. Connection point 2: Conductive parts, protective earth and auxiliary ports complying with the limits of SELV shorted together.	Separation between interface I_a and user accessible ports	Test voltage: 1000 V a.c. or 1414 V d.c. Test time: 2 s min, 6 s max	No breakdown	10, 11, 12–15
7	Where acoustic shock protection relies on specific components and possibly their correct orientation then the integrity of these circuits should be verified.	Acoustic shock	BS 6450: Part 2: 1983: Clause 6.2.10 or TBR8: Annex C or BS 6317: 1982: Clause 13.9 or 85/013: Issue 4: Clause 5.2.9	+224 dBPa	14

Notes on BABT electrical and acoustic safety tests:

1. The test as described is for Class I equipment, i.e. equipment that relies on a protective earth connection for providing safety. (This test is in accordance with BS EN 60950, Clause 2.5.11.)

2. As an alternative location for connection point 1, for equipment incorporating a mains supply cord the supply earth connection (normally the earth pin of the mains plug) of the cord should be used. In this case the measured resistance should be not greater than $(0.1 + R)\ \Omega$, where R is the resistance of the earth lead within the mains supply cord.

3. On equipment where the protective earth connection to a sub-assembly or to a separate unit is by means of one core of a multicore cable that also supplies mains power to that sub-assembly or unit, the resistance of the protective conductor in that cable should not be included in the resistance measurement (as is the case with the mains cord resistance, see Note 2). Where the cable is protected by a suitably rated protective device (which takes into account the impedance of the cable), the minimum test current may be reduced to 1.5 times the rating of this protective device.

4. The test as described is for Class I equipment. Normally the protective earth connection is the earth pin of the mains plug.

5. This test is applicable to Network Terminating Points (NTPs) connecting to either analogue or digital networks, but see Note 11.

6. Where the user accessible conductive part, or the NTP connection, is isolated from primary circuits by reinforced or double insulation but is either
   i   connected to protective earth for functional reasons, or
   ii separated from protective earth by less than supplementary insulation,
   then it may not be possible to conduct this test without overstressing basic insulation (which is only designed to withstand 1500 V a.c. (for further explanation see the relevant notes to BS EN 60950, Clause 5.3.2). In this case, the individual components providing the reinforced or double insulation should be tested in accordance with test No. 5, but the finished equipment may be tested in accordance with test No. 2. When testing such finished equipment, any user accessible conductive parts that are not connected to Earth in the equipment should be connected to protective earth when performing test No. 2.

7. The test voltage specified may be increased at the discretion of the manufacturer. However, BABT do not require any higher test voltages.

8. The pass/fail general criterion for electric strength tests is that no breakdown should occur. In practice, the trip current on the test instrument will need to be set so that it does not trip when subject to the normal leakage (touch) current (predominant for a.c. testing) or insulation resistance current for the equipment and test voltage concerned. Trip current levels should be set to a minimum practical level.

9. For an item of equipment supplied with a UK 13 A mains plug having a leakage of 3.5 mA or less, the following maximum trip currents are acceptable in accordance with BS EN 60950. For hand-held Class II equipment lower trip currents are appropriate. See BS EN 60950, Clause 5.2.2.

Test voltage	1500 V a.c.	3000 V a.c.	2121 V d.c.	4242 V d.c.
Trip current	42 mA	84 mA	1.5 mA	3.0 mA

10. The 2 s for the test duration should apply where operator judgement against a time standard is used. Where the timing is carried out automatically, the minimum test time may be reduced to 1 s (the 1 s minimum is in accordance with the note regarding production tests in BS EN 60950, Clause 5.3.2). The 6 s maximum is advisory. However, BABT deprecate longer production test times and repeated electrical strength testing as they can cause damage to insulation.

11. This test is applied only to Digital Terminating Equipment as an option to test No 3.

12. Where appropriate, the sequence of safety tests should be earth continuity followed by electric strength.

13. Where the integrity of each test may be demonstrated, it is possible to combine certain of these tests.

14. Where tests on sub-assemblies have been conducted under BABT surveillance, it is not necessary to re-test the complete assembly if overall compliance with the appropriate tests is demonstrated to BABT and if final assembly arrangements do not affect the integrity of the sub-assembly tests.

15. When the above tests are to be applied to power supply units the low voltage outputs are considered to be the NTPs.

Model forms for in-service inspection and testing

Form VI.1   Equipment register

Organisation			Address		
Responsible person					

Register no.	Location	Equipment description	Serial no.	Frequency of	
				formal visual inspection	combined inspection and test

Date:					Page ☐ of ☐

Notes on the formal visual and combined inspection and test record (Form VI.2):

1 Register No. – this is an individual number taken from the equipment register for this particular item of equipment.
2 Description of equipment, e.g. lawnmower, computer monitor.
3 Construction Class. Class 0, 0I, I, II, III. Note that only Class I and Class II equipment may be used without special precautions being taken.
4 Equipment types – portable, movable, hand-held, stationary, fixed, built-in.
5,6 Insert the location and any particular external influences such as heat, damp, corrosive, vibration.
7,8 Frequency of inspection – generally as suggested in Table 7.1 of this Code of Practice.

   Inspection. Items 17–23 and 28–31 will be completed if an inspection is being carried out.
   Inspection and test. The testing in items 24 and 26 should always be preceded by inspection.

9–11 The make, model and serial number of the item of equipment should be inserted.
12–14 The voltage for which the equipment is suitable, the current consumed and the fuse rating should be inserted.
15,16 The date of purchase and the guarantee should be completed by the client.
17 The date to be inserted is the date of the inspection or the date of the inspection and testing.
18 Environment and use. It should be confirmed that the equipment is suitable for use in the particular environment and is suitable for the use to which it is being put.
19 Authority is required from the user to disconnect equipment such as computers and telecom equipment where unauthorized disconnection could result in loss of data. Authority should also be obtained if such equipment is to be subjected to the insulation resistance and electric strength tests.
20 Socket-outlet/flex outlet. The socket-outlet or flex outlet should be inspected for damage including overheating. If there are signs of overheating of the plug or socket-outlet, the socket-outlet connections should be checked as well as the plug. This work should only be carried out by an electrician.
21–23 The inspection required is described in Chapter 14 of this Code of Practice.
24–27 Tests. The tests are described in Chapter 15 of this Code of Practice. The tests should always be preceded by the inspection in items 11–17. The instrument reading is to be recorded and a tick entered if the test results are satisfactory.
28–31 These columns are to be completed when inspection is performed and when inspection and testing are performed.
28 Functional check. A check is made to ensure that the equipment works properly.
29 Comments/other tests. Additional tests may be needed to identify a failure more clearly or other tests may be carried out such as a touch current measurement. An additional sheet may be necessary, which should be referenced in the box on this record.
30 OK to use. YES should be inserted if the item of equipment is satisfactory for use, NO if it is not.

Form VI.2 Equipment formal visual and combined inspection and test record

Item of equipment	1	2	3	4	5
[1] Register no.					
[2] Description					
[3] Construction class					
[4] Equipment type (P, M, HH etc.)					
[5] Location					
[6] Particular requirements of location					
[7] Frequency of formal visual inspection					
[8] Frequency of combined inspection and testing					
[9] Make					
[10] Model					
[11] Serial no.					
[12] Voltage (V)					
[13] Current (A)					
[14] Fuse (A)					
[15] Date of purchase					
[16] Guarantee					
[17] Date					
[18] Environment					
[19] Disconnected					
[20] Socket-outlet					
[21] Plug					
[22] Flex					
[23] Body					
[24] Continuity (Ω)					
[25] ✓					
[26] Insulation (MΩ)					
[27] ✓					
[28] Functional check					
[29] Comments					
[30] OK to use					
Initials					

Note: (✓) Indicates pass, (x) Indicates fail, (N/A) Not applicable, (N/C) Not checked

Form VI.3   Equipment labels

Date of check _____

Initials _____

Appliance no. _____

Next test before_____

PASS

✓

SAFETY CHECK

**DANGER
DO NOT USE**

Date of check _____

Initials _____

Appliance no. _____

FAIL

✗

SAFETY CHECK

Form VI.4   Repair register

Organisation			Address				
Responsible person							

Register no.	Customer	Description	Serial no.	Repairer	Suitable for return to use		
					✓	Signature	Date

(✓) Indicates satisfactory   (x) Indicates unsatisfactory

Form VI.5   Faulty equipment register

Organisation			Address	
Responsible person				

Date	Register no.	Equipment fault	Location	Action taken

Code of Practice for In-Service Inspection and Testing of Electrical Equipment
© The Institution of Engineering and Technology

Form VI.6   Test instrument record

Organisation	Address
Responsible person	

Testing points for low resistance and high resistance tests

Low resistance

Type:	Model:		Serial no.:		Date of last calibration:	
Date of test						
0.5 Ω						
Deviation ± %						
Date of test						
1.0 Ω						
Deviation ± %						
Date of test						
10.0 Ω						
Deviation ± %						

High resistance

Type:	Model:		Serial no.:		Date of last calibration:	
Date of test						
0.5 MΩ						
Deviation ± %						
Date of test						
1.0 MΩ						
Deviation ± %						
Date of test						
10.0 MΩ						
Deviation ± %						

Other

Type:	Model:		Serial no.:		Date of last calibration:	
Date of test						

Resistances of flexible cords

Table VII.1 gives figures for the nominal resistance of the protective conductor per metre length and for various lengths of cable that may be fitted as supply leads to appliances. Once an earth continuity test has been performed the approximate resistance of the protective conductor can be found and deducted from the test result to give an accurate figure for the earth continuity reading of the appliance (see Section 15.4).

▼ **Table VII.1** Nominal resistances of appliance supply cable protective conductors (figures are for cables to BS 6500 or BS 6360)

Nominal conductor csa	Nominal conductor resistance at 20 °C	Length		Resistance at 20 °C		Maximum current-carrying capacity	Maximum diameter of individual wires in conductor	Approx. no. of wires in conductor
(mm²)	(mΩ/m)	(m)		(mΩ)		(A)	(mm)	
0.5	39	1	1.5	39	58.5	3	0.21	16
		2	2.5	78	97.5			
		3		117				
		4		156				
		5		195				
0.75	26	1	1.5	26	39	6	0.21	24
		2	2.5	52	65			
		3		78				
		4		104				
		5		130				
1.0	19.5	1	1.5	19.5	29.3	10	0.21	32
		2	2.5	39	48.8			
		3		58.5				
		4		78				
		5		97.5				
1.25	15.6	1	1.5	15.6	23.4	13	0.21	40
		2	2.5	31.2	39			
		3		46.8				
		4		62.4				
		5		78				
1.5	13.3	1	1.5	13.3	20	15	0.26	30
		2	2.5	26.6	33.3			
		3		39.9				
		4		53.2				
		5		66.5				
2.5	8	1	1.5	8	12	20	0.26	50
		2	2.5	16	20			
		3		24				
		4		32				
		5		40				
4	5	1	1.5	5	7.5	25	0.31	53
		2	2.5	10	12.5			
		3		15				
		4		20				
		5		25				

Note: 1000 milliohms (mΩ) = 1 ohm (Ω)

Code of Practice for In-Service Inspection and Testing of Electrical Equipment
© The Institution of Engineering and Technology

Checks to be made on a plug, a cable and an extension lead

1	Check that the plug is suitable for the application. A resilient plug marked BS 1363A may be necessary if the plug is subjected to harsh treatment, e.g. vacuum cleaners, lawnmowers and extension leads.
2	A plug with a cracked or otherwise damaged body should be replaced. 
3	Look for signs of overheating. This may be caused by a fault in the plug such as a loose connection or by a faulty socket-outlet. It can also be caused by a loose connection where the fixed wiring is connected to the back of the socket-outlet. The overheated plug and fuse should be replaced. It is very likely that the socket-outlet into which it has been plugged is also damaged and should be replaced. 
4	Check that the flexible cable is properly secured in the cord anchorage and positively gripping the sheath. There should be no strain on the cable cores or the terminations. 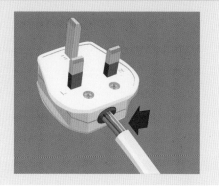

continues

Code of Practice for In-Service Inspection and Testing of Electrical Equipment | **125**
© The Institution of Engineering and Technology

▼ **Table VIII.1**
continued

5	If the plug is of the non-rewirable or moulded-on type, the cable grip should be tested by firmly pulling and twisting the cable. No movement should be apparent.	
6	The plug has to be correctly connected. The brown core has to be connected to the phase or L (live) pin, the blue core has to be connected to the N (neutral) pin and green-and-yellow has to be connected to the E (earth) pin.	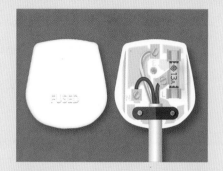
7	Check, with a screwdriver, that the cable core terminations are tight and secure. The missing screw illustrated in the figure means an intermittent and unreliable connection with Earth.	
8	An excessive amount of insulation should not have been removed. In this example, too much insulation has been stripped off the cable cores. This could pose a risk of short circuit.	
9	There should be no loose strands. Loose strands mean there will be a poor connection between the conductor and the plug pin with a risk of overheating. In addition there is a risk of a loose strand making a short circuit.	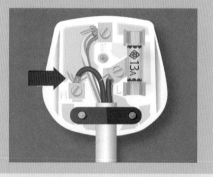
10	Check that the cable cores in the plug are not strained.	

11	The fuse should be securely gripped, and should not show any signs of overheating. Check that the fuse is to BS 1362 and is approved. An ASTA mark shows that it has been approved for safety.	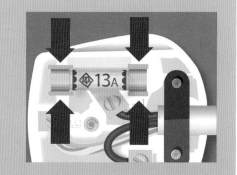

Check the rating of the fuse. The manufacturer's recommendations should be followed. If the manufacturer has fitted a particular fuse, the rating of the fuse should not be changed. Most appliances up to about 700 W should have a 3 A fuse fitted (red). Appliances over about 700 W should have a 13 A fuse (brown) fitted. However, account should be taken of start up currents such as motor starting currents and transformer inrush currents. Note that certain IT equipment is fitted with a 5 A fuse and this fuse should not be replaced by one of a different value.

Non-rewireable plugs will have the appropriate fuse rating marked on them.

12	A BS 1362 fuse should be fitted. The foil shown in the diagram will not provide any electrical protection and will probably make a poor connection resulting in overheating with a risk of fire.	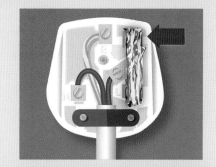

13	A loose or disconnected circuit protective conductor means that where an item of Class I equipment is supplied the equipment is not earthed. This is a direct contravention of Regulation 8 of the Electricity at Work Regulations. Regulation 8 is an absolute regulation meaning that its requirements have to be met regardless of cost or difficulty.	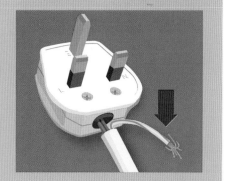

continues

▼ **Table VIII.1**
continued

14	If the label is not removed, the plug may not make a good connection in the socket-outlet and overheating may result.	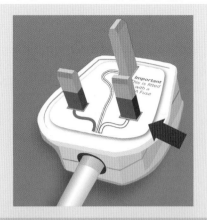
15	All standard 13 A plugs now sold are required by law (The Plugs and Sockets etc. (Safety) Regulations 1994) to conform to BS 1363, which requires pins to be sleeved. The legislation is not retrospective in that it does not apply to old plugs already in use, but such plugs should not be reused (i.e. not refitted to new cords).	
16	When replacing the plug cover check that it fits properly and will not come loose during use.	

Note: Some of the above checks may not be possible for equipment fitted with a non-rewireable or moulded plug.

▼ **Table VIII.2**
Checks to be made on a cable

1	Is the cable too long or too short?
2	Check that a detachable power supply cord supplying an item of Class I equipment incorporates a continuous protective conductor.
3	Check the flexible cable connections and anchorage at the equipment, if practicable.
4	Having removed the plug or otherwise switched off the supply, inspect the flexible cable for damage throughout its length both by visual inspection and touch. Supply cords should not be extended by taped joints and should not exceed the length allowed by the equipment standard. Where extension leads are necessarily used, the guidance given in Section 15.10 should be followed. Cables should not be bent into tight loops. Equipment should not be positioned so close to walls or skirting boards that cables are forced into tight bends either where they enter the equipment or where they exit the plug.	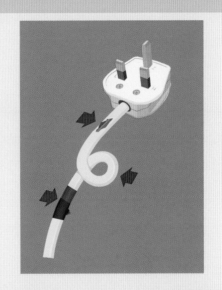
5	Damaged cable should always be replaced or removed by cutting out the damaged part and refitting the plug.	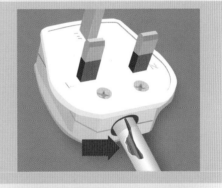
6	Non-flexible cable, such as twin-and-earth, should never be used for connecting an appliance or for an extension lead – the regular bending in use will damage the cable.	

continues

▼ **Table VIII.2**
continued

7	A 13A plug is designed for one cable only – not two.	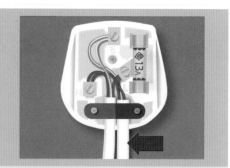
8	The cable should include a protective conductor if an item of Class I equipment is being supplied. All extension leads should include a protective conductor.	
9	Supply cords should not exceed the length allowed by the equipment standard.	
10	Extension leads fitted with an RCD should be checked and tested as indicated in Section 15.10.	

▼ **Table VIII.3** Checks to be made on an extension lead

1	Any extension lead that does not have a protective conductor (or earth wire) should be marked as defective and removed from service. Extension leads fitted with 3-pin plugs and sockets should never be wired with two-core cables as there will always be the possibility that the lead will be inadvertently used to supply a Class I appliance, which as a consequence, would not be earthed.
2	Cable reels should be used within their coiled or uncoiled ratings as appropriate because an extension lead can generate considerable heat when heavily loaded. 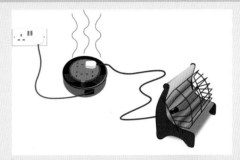
3	A cable reel should not be used outdoors unless designed for such use. 
4	An extension lead should not be run under a carpet as the carpet will impede heat dissipation possibly causing the cable to overheat.
5	Check that an extension lead is not a trip hazard.
6	An extension lead should never be made using non-flexible cable such as flat twin-and-earth as the flexing of the cable will result in failure. Any such extension lead should immediately be removed from service.
7	Equipment should only be supplied by means of an extension lead on a temporary basis. Extension leads should not be used on a permanent basis. BS 7671 requires that a sufficient number of conveniently placed socket-outlets be provided.

8	The length of an extension lead should not exceed that given in the table on the right.	**Maximum lengths of an extension lead**	
		csa of the cores (mm²)	Maximum length (m)
		1.25	12
		1.5	15
		2.5	25

9	Where an extension lead includes an RCD, the RCD must have a rated residual operating current not exceeding 30 mA.
	The RCD should be checked for correct operation by plugging it in, switching it on and then pushing the test button. The RCD should operate and disconnect the supply from the socket(s) fitted on the extension lead.

For further information refer to the HSE publication HSG 85: *Electricity at Work: safe working practices*.

Guide to isolation procedures

1   Verify that it is both safe to perform the isolation and acceptable with the occupier or user.

2   Ensure that the circuit, its conductors and the disconnector have been correctly identified.

3   If the disconnector is an off-load device, switch off the load.

4   Open the disconnector and secure it in the open position with a lock or other suitable means.

5   Prove the correct operation of a suitable voltage detection instrument against a known voltage source.

6   Using the voltage detector, check that there are no dangerous voltages present on any of the circuit conductors to be worked on. Note that conductors may be energized due to a circuit fault or incorrect wiring such as a borrowed neutral. Check circuit protective conductors also, as these may be incorrectly connected.

7   Prove the voltage detection instrument against the known source to establish that it was functioning correctly when the conductors were tested.

8   Ensure the supply cannot be inadvertently or unintentionally reconnected. The supply must remain disconnected by means such as locking the isolator or circuit-breaker or removal of fuses. In addition, labels should be used to warn persons.

Index

Index

Code of Practice for In-Service Inspection and Testing of Electrical Equipment
© The Institution of Engineering and Technology